750

75

Eye color, sex,
and
children's behavior

Eye color, sex, and children's behavior

A. L. Gary, Ed. D.
John Glover, Ed. D.

Nelson-Hall Publishers, Chicago

Gary, A   L   1935-
    Eye color, sex, and children's behavior

    Bibliography: p.161
    Includes indexes
    1. Genetic psychology.   2. Child psychology.
3. Color of eyes—Psychological aspects.   4. Sex
differences (Psychology)   I. Glover, John A., 1949-
joint author.   II. Title. [DNLM: 1. Eye color.
2. Child behavior.   WS105 G246e]
BF 701.G29   155.2'34   76-3642
ISBN 0-88229-213-7

Contents

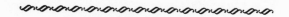

-

Dedication

The dedication of this text is not made to an individual or group of individuals, but to an idea—a value. The authors are committed to the value of individual differences and human diversity, with the sincere belief that the nature of humanity is so diverse (and is so with good cause—that of survival) that it is absurd to attempt to force all people into a single, evaluative mold, to respond to all individuals by one procedure or process, or to hold up the "holy mean" as the only standard by which to consider the worth of our fellowmen.

Preface

Adequate descriptions, explanations, and accurate predictions of the behavior of any organism are concerned with three main factors: the state of the organism, its history of reinforcement, and the current conditions under which the behavior under consideration is occurring.

Lorenz (Travis, 1974), a leading ethologist, has defied the Skinnerians to condition a female rat or a female pigeon to lie on her back during copulation, or to settle a greylag goose colony in a different location. Skinner (1974), in addressing the same issue, said that in one sense, all behavior is inherited, since it is a product of natural selection. In the same text, Skinner recognizes the importance of physiological research. He states that the behavior of an organism will eventually be said to be determined by what it is at the moment it behaves and how it came to be so—both as an individual and as a member of the species—and that the physiologist will finally give us all the details.

Both scientists were addressing the problem of finally determining the "causes" of behavior. It has appeared that the further up the phylogenetic scale an organism is located, the less important the state of the organism is in the determination of behavior. This opinion is due, perhaps, to the profound changes that can be brought about in human behavior by the manipulation of the environment. The changes that are possible are so dramatic that they have overshadowed somewhat the factors of history of reinforcement and in particular the state of the organism.

Another problem behavioral scientists have faced in the determination of causes of behavior is that the conditions affecting behavior are *most accessible* when considering the factor of current conditions; they are *much less accessible* when considering the factor of history of reinforcement; and they are *virtually out of reach* when we consider the state of the organism factor. This fact is most evident when we review the nomothetic research on the "causes" of behavior. There is very often a huge unexplained variance in scores called the error term which cannot be attributed to any of the independent variables. Behavioral scientists blithely explain this variance as being attributable to the organisms' history of reinforcement and the state of the organism (including inheritance). These research findings rudely remind us as behavioral scientists of the degree of refinement of our "science". The problem, therefore, is a two fold one: first, if we are to have a refined science, we must be able to compartmentalize history of reinforcement and state of the organism as general classes of factors affecting behavior; second, we must be able to stipulate with much more precision the specifics within each of the compartments.

To return to the effects of the research on operant conditioning (manipulating the environmental conditions), as researchers we perhaps have been deluded into not attending to the questions of the state of the organism in our quests to determine the causes of human behavior by virtue of the fact that current environmental conditions and history of reinforcement *appear* to be much more important. The word *appear* is used here for two reasons: (1) a much greater proportion of lower animals' behavior is believed to be attributable to inherited factors, and (2) the increasing inaccessibility of the contingencies of behavior as we move from higher to lower organisms. Current conditions involve primarily the contingencies of reinforcement; history of reinforcement is concerned with both the contingencies of reinforcement and the contingencies of survival; the state of the organism is concerned primarily with the contingencies of survival. What we are concerned with, then, is separating the contingencies of reinforcement from the contingencies of survival—and one may readily see from the discussion in the previous paragraph that the contingencies of survival, the state of the organism, and the factors which affect behavior and which are virtually inaccessible are probably one and the same.

This text, in dealing with eye color—an inherited characteristic that is linked to survival behaviors in lower animals and to general response forms or modes in humans—attempts to shed some light on how we might proceed to discover more about the contingencies of survival in the behavior of *homo sapiens,* and in doing so, make future analyses of human behavior more accurate and powerful. It may be that the contingencies of survival are much more significant to the explanation of human behavior than we care to accept.

Chapter 1 of the text is concerned mainly with the foundations upon which the balance of the text is built. It describes in a general way the findings of Dr. Morgan Worthy, the pioneer researcher in the area of eye pigmentation and human behavior. Chapter 2 presents a study and a discussion of eye color and kinesics. Chapters 3 and 4 present research on creativity and its relationship to eye color, and reactive and self-paced behavior. Modeling, an extremely important aspect of behavior, is covered in chapter 5. People of different eye shades perceive and react to color and form differently; chapter 6 presents a study and a discussion on this phenomenon.

Chapters 7, 8, and 9 cover matters such as eye color and sociability, some "pilot study" information on eye color and addictive behavior, and the educational implications of the eye color finding.

Throughout the text, the authors have used the Worthy technique for rating eye color. Judges were asked to rate eye colors from 1 through 5, with 1 representing light blue and gray and 5 representing dark brown. The scale midpoint is green or hazel. Colors 1 through 3 are collapsed to form the light-eyed grouping; colors 4 and 5 make up the dark-eyed groups. This information is presented here in the interest of clarity and parsimonious discourse in the text.

Acknowledgments

Grateful recognition is extended Dr. Morgan Worthy of the Counseling Center, Georgia State University, Atlanta, Georgia, for his pioneer work in the development of the concepts elaborated upon in this text. Without his creative thinking and teaching skills, and without his scientific perseverance in the systematic examination of data from a number of disciplines, this book could never have materialized. The works contained herein were inspired by the stimulating and provocative *Eye Color, Sex and Race: Keys to Human and Animal Behavior* (1974).

1
Foundations
and
preview

I n his book, *Eye Color, Sex and Race: Keys to Human and Animal Behavior* (1974), Dr. Morgan Worthy presents an overwhelming amount of evidence in support of a new perspective in the examination of the nature of human and animal behavior. His text is particularly strong because much of the data were collected by other scientists representing other disciplines, and before his hypotheses were ever formulated. The data collected by others and used by Worthy were gathered in the study of matters unrelated to Worthy's ideas. The independent nature of that evidence makes it remarkably bias free, a research condition every scientist strives for. In the design and execution of his own experiments, Worthy followed the leads of his personal speculations and the avenues that opened as a result of the findings of others, regardless of where those leads and avenues took him. The result is a highly refined system of interlocking concepts that raises more questions and opens more lines of inquiry than it provides answers, and these are the benchmarks of good research. The heuristic value of a given line of inquiry is one of the main criteria by which fruitful research is evaluated.

As a prelude to the presentation of the follow-up information on eye color and behavior in this volume, it is necessary to set forth the developmental process of Worthy's research, the concepts he formulated, and some of his operational definitions that will set the stage for further investigations into the assumptions underlying his work. Then a series of chapters will follow, each of which views ideas stimulated by his work through the presentation of original research projects conducted by the authors. The book will conclude with a chapter on further speculations and will suggest research designs for further investigation.

The activities of the three-plus years that went into the writing of Worthy's book were initiated by his observation that there seemed to be differences in sports performances between black and white athletes, and that these differences were somehow tied to the type of activity demanded for success by the particular sport or sport position under consideration, such as quarterback, golfer, boxer, and so on. Worthy pursued his hypothesis and reported his findings in a scientific paper (Worthy & Markle, 1970) which reported statistically significant differences between black and white sports performances on a number of dependent measures. Rather than continue to follow a line of reasoning that had already been shown to be faulty and one that conflicted with his personal value system, Worthy concentrated on individual differences, but he also chose to watch for clues that would indicate that there are reasons other than skin color for performance differences.

Probably two of the least explored and unheralded components of "raw" intelligence are the ability to recognize the "unusual and the unique" and to respond to them as such, and to recognize as extraordinary those events and characteristics that seem to most people to be truly mundane. Worthy made such observations, and although his "common sense" told him (along with a lot of people) that his observations were not important, his creative and scientific intelligence told him otherwise. The observation was that there seemed to be individual differences in performances of certain types of behaviors that were highly correlated with eye color—a truly absurd finding that would not seem worthwhile for any serious scientist to pursue. But Worthy had a "hunch," and he followed up on it. When he examined differences in performance within the groups of white athletes, he found that there were, indeed, statistically significant differences by eye color. This provided Worthy the opportunity he was looking for: a chance to explore the nature of the differences in human behavior, but in relation to a factor other than skin color. Since the factor of eye color is inherited, it allowed him to look at the behavior "styles" that he had by that time operationally defined and to consider them from an evolutionary standpoint. Particularly, he could review the behavior of animal groups and look for consistent patterns up and down the phylogenetic scale. This is precisely what he did, and the consistencies were found. Encouraged and excited by his findings, Worthy then moved back and forth between human and animal research as he made careful comparisons on psychobiological dimensions. He compared speed of locomotion, survival behaviors, and the functions and characteristics of the eye across humans and animals. They all seemed to be related to one another in a systematic way. During the first year or so of research, Worthy formulated a number of constructs and operational definitions as he began to develop a system and order to his inquiry.

The two main behavior constructs upon which his investigations rest are the two general categories or "styles" of responses found in both humans and animals, both of which are related to eye pigmentation. Worthy named the two constructs "self-paced" and "reactive" behaviors, and operationally defined them thus: self-paced behavior is that which occurs when the stimulus conditions are relatively fixed and when the organism has rather broad time limits in which to respond; reactive behavior is that which occurs when the stimulus conditions are rapidly changing and when the organism must respond quickly for success. There are two elements that are crucial in the definitions: stimulus conditions and response time. Worthy points out that there are two things critical to understanding reactive behavior: (1) the organism cannot plan a specific response in advance; one may plan for contingencies of response for situations often encountered, but if the variety of situations is great, success will often depend upon spontaneity; (2) the organism has little or no control over the time of the response; one cannot wait to "get ready" nor unduly inhibit a response. Worthy often offers sports activities as examples of his two behavior constructs. A self-paced sport is golf; a reactive sport is boxing. Of course, there are many sports having several team members in which the performance requirements of the positions vary from that of being largely reactive and somewhat self-paced to somewhat reactive and largely self-paced. A football team is such an example.

The quarterback's position requires very skillful planning and execution; being able to inhibit a response is critical: not releasing the pass until the proper instant, "checking off" a call when the defense has realigned in anticipation of the play that is being called, and so forth. The pass receiver must be highly responsive to the defensive backs' moves. Although the end has a pattern to run, he must very often deviate from his assignment to be successful. His responses are triggered by unanticipated adjustments on the part of the defense.

After the discovery of the eye color relationship to both human and animal behavior, Worthy moved largely to the study of animals. In doing so, he used findings from several disciplines, including anthropology and biology, and used primate and ornithological studies, among others. He picked studies in which eye color was included as a descriptive characteristic but was not necessarily controlled for in an experimental sense (not treated as an independent variable). During the course of collecting the studies, two important things were noted: (1) in many species, a wide range of eye color was found within the species, and (2) many of the behaviors recorded by other scientists and reanalyzed by Worthy were survival behaviors, that is, related to the acquisition of the animal's chief food, mating, fight and flight skills. Both made the evolutionary argument—species survival and natural selection—very strong.

The environment is constantly changing due to shifting animal populations and shifting supplies of food—both of the animal and vegetable type. Climatic changes occur, both routine (seasonal) and for other, more complex geological and meteorological reasons. To survive, a species must have adaptability within the individual, within the species, or both. If the adaptability is specific to some individuals and not to others, ultimately (perhaps) the individuals who are most adaptable survive. It is postulated that the reason a *range* of eye color is found in many species, including *Homo sapiens,* is that eye pigmentation is a visible indicator of a range of responses (adaptability) that is *species-specific.* For instance, if a species has a rather wide range of eye color represented in its population, and the predominant eye color is dark, for example, then acquisition of the chief food would normally involve highly reactive skills. But presume, for purposes of explaining the "survival of the species" proposition, that the environment quickly changes and the chief food of the species suddenly disappears with the only foods left requiring self-paced behaviors for acquisition. The chances are that the lighter eyed of the species would survive by virtue of having some capability of response inhibition. The light-eyed survivors would reproduce, but among their offspring would be some dark-eyed because "dark eyes" were in the gene pool. The within-species range of eye color characteristic is likely a precaution against species extinction, and perhaps is nature's way of guarding against drastic environmental changes being species-fatal.

In reviewing animal behavior, Worthy noted that

there are two types of survival behaviors that are analogous to self-paced and reactive human behaviors, and that they are highly correlated with eye color. The dark-eyed animals he researched seem to have a combination react-approach-flee set; light-eyed animals seem to have a combination wait-freeze-stalk set. The fact that these behaviors are linked to survival and to eye-color and are comparable to the reactive/self-paced continuum used in the description of human behavior makes the evolutionary aspect of the argument very strong, indeed.

Worthy then delved into the functions of the eye—both in humans and along the phylogenetic scale. In a rather technical chapter he discusses the competing functions of the eye, sensitivity and resolving power, both of which are associated with eye color. He explains the relationship between eye color and color versus form perception through a number of experiments done by other scientists, and shows how eye color, competing functions of the eye, and color versus form perception are related. The findings he presents are extremely import- ant because of their significance to the understanding of human responses to and interpretation of environmental stimuli. Not only is eye color linked to the response speed of an organism, but it also determines the aspects of the environment that are salient to the organism's perception.

Another issue raised by the work of Worthy is of primary concern to psychologists and other professional people interested in the composition of research designs. In the group analysis of human behavior, researchers gather information about their subjects that can be expressed along one or more dimensions. For example, in a one-way analysis of variance research design, there is one independent variable and one dependent variable. Male subjects might be divided into two or more groups based on their ages, such as ages sixteen through eighteen, nineteen through twenty-one, and twenty-two through twenty-four. The independent variable is "males," and each of the three age groupings is a level of the independent variable. The dependent measure, or the "thing" we will measure in our hypothetical design, might be the distance each of the subjects can throw a baseball. In such a design, each male would have a score that would probably be expressed in feet and inches. The scores of the three age groups of males would be collected and analyzed by the one-way analysis of variance technique to determine if there were significant differences between the three male groups on the dependent variable. If the design were actually carried out, the results obtained would not be very valuable; the reason being that so many factors already known to affect an individual's score are not controlled for in the design. Elements such as height, weight, general condition of health, and prior experience throwing a baseball could be confounded in the results of the experiment. The point of the example cited has to do with the factors that make up the total score of a given individual in a psychological experiment. Each person's score is made of many components; the nature of the design sets the limit on how many sources of the score can be properly identified. In the example furnished, each of the scores could be broken down into only two parts. One portion of each score would be attributable to being "male" and to being a member of a particular age group. The second, and perhaps the larger component, would be accounted for statistically as "unexplained variation" or "error term." The larger the error term, in general, the less valuable or less conclusive the results of the experiment. Therefore, the task of the psychologist is to examine human behavior while controlling for the many factors that contribute independent or somewhat overlapping parts to the total score of each individual. As the number of factors that contribute to each person's score are increased in the design, the error term is decreased. Said another way, the smaller the unexplained variation in scores, the more precise and more certain one can be of the conclusions reached from the data. The contribution that Worthy's work has made to the statistical analysis of human behavior in nomothetic designs is in the area of error term reduction. Many times, by executing designs that control for eye color, a greater portion of each individual's score is accounted for, thereby adding to the precision and power of the design. While considering the theoretical contributions to research designs made by eye color blocking, it might be appropriate to mention that many published research articles that have not controlled for eye color could perhaps be replicated, but modified to the extent that eye color is added to the independent variables. As will be shown in subsequent chapters, eye color sometimes contributes more to changes in scores in pre- and postmeasure designs than does the experimental treatment. This is an important effect, indeed, and should not be taken lightly in future research.

While on the topic of research design and blocking, it should be mentioned that sex has been effectively used as a blocking variable to increase the efficiency of research designs, that is, to make them more powerful. Studies are cited by Worthy in which both sex and eye color are linked to self-paced and reactive behavior. In some designs, both sex and eye color interact powerfully with the treatment dimensions. Interactions, when they are produced in multifactor designs, are often more interesting and more fruitful scientifically than main effect findings.

In summary, Worthy's findings are that dark-eyed organisms tend to be more reactive and light-eyed organisms are more self-paced. Another way of stating the above is that dark-eyed organisms do not have the ability to inhibit a response as well as their light-eyed counterparts. Further, this holds true for organisms all up and down the phylogenetic scale.

On one level, eye color seems to be very important in that the amount of light that enters the eye is governed by iris pigmentation, an inherited characteristic. The amount of light entering the eye seems to be directly related to the activation level of the organism. In addition to being associated with the activation level of the organism (response), it seems also to be related to perception in terms of the resolving power and sensitivity continuum.

But there is a deeper level in the investigation of eye color and behavior that is outside the research expertise of the authors of this book. Eye color is genetically determined; however, some genes are not selected singly, but in clusters. Therefore the notion that eye color—in and of itself—is important in behavior analysis may be a rather superficial one. It may be that it is the genes selected along with the eye color genes, or the characteristic eye color and gene groupings, that are important.

What, exactly, do Worthy's findings mean? The evidence may be summarized as follows: An inherited trait, eye color, is highly correlated with survival behavior repertoires in animals. Further, when certain human behaviors are analyzed, they seem to fit the response modes of the animals; that is, it seems plausible to analyze them in terms of the self-paced/reactive continuum, and eye color is also correlated with response style in certain human behavior. Statistical analysis tells us that the correlation between the eye color and the behavioral style phenomenon is not a chance happening. How, then, do we explain the findings? The answer is that we are not at all certain. One hypothesis is that the amount of light entering the eye determines the amount of epinephrine that is released into the bloodstream. If organisms of certain eye colors have more ephinephrine released onto the bloodstream (particularly under stress conditions) than organisms of other eye colors, these organisms should be less able to inhibit a response. Hence, the "reactive" style of the darker eyed organisms. Since darker eyed organisms, including humans, have obviously survived, it seems that the natural selection process has reinforced the reactive style. Stated another way, if darker eyed organisms do not as a rule inhibit responses as much as light-eyed organisms, it is reasonable that one would be much more likely to be reinforced for successful uninhibited responses (one cannot be reinforced for a behavior one does not emit). The same is true for self-

paced behaviors and light-eyed organisms. If one's ana-
tomical structure permits only a low frequency of uninhibited responses and a high frequency of inhibited
responses, it is likely that more reinforcement will accrue
to the high probability response mode.

It is very possible that eye color and the release of
epinephrine factors are only part of the explanation.
Since eye color is genetically determined, and since some
genes are not selected singly but in clusters, it is possible
that a large number of organ structures and functions are
different in dark-eyed and light-eyed organisms, and
these differences may just "happen" to be correlated with
eye color.

Consideration of the complexity of the world in
which we live and the wide range of behaviors exhibited
by people as they engage in success behaviors gives rise
to the thought that we perhaps have spent too much
effort in the rank-ordering of large numbers of people on
very narrow standards of performance. The survival of
the species issue raised by Worthy in relation to the eye
color and behavior type of organisms suggests that we
move away from valuing the norm and more to valuing
diversity; our very survival may depend upon it. This
may take some mental gymnastics on our part, for we
have learned to do (or have been reinforced for doing)
just the opposite.

The chapters to follow deal with specific areas of
interest to the authors: that of eye color and classroom
behavior. A chapter will be presented on eye color, sex,
and kinesics. Kinesics is an area of investigation that is
capturing more and more of the time of serious behavioral scientists. The kinesic findings strongly suggest
that the nonverbal portion of communication may actually contain more information than the verbal portion. If
this is true, future kinesic studies should certainly
control for eye color. And certainly, interpersonal communication is important in the classroom—it is why
people are there in the first place.

Creativity in relation to eye color and sex, along with

a learning technique called modeling, will also be the subject of a chapter. As in the kinesics chapter, an original study will test Worthy's behavioral reactivity hypothesis in an interesting design.

Milton Rokeach developed a scale which measures an attitudinal construct he calls "Dogmatism." Eye color and sex are controlled in a 2 x 2 design using high school students' Dogmatism scores as the dependent variable. The Worthy hypothesis is tested to determine if behavioral responsivity, eye color, and sex are related to the attitude called Dogmatism.

Worthy's complicated and technical section on the competing functions of the eye and on the phenomenon of color versus form perception that is associated with eye pigmentation is tested to see if it holds for a special population of secondary public school children.

In another original study carried out in classrooms of the public schools, the behavioral reactivity hypothesis is tested regarding student response speed to questions on academic subject matter in an "open answer" situation. It is suggested that the behavioral reactivity research may shed some light on the diagnostic category called "the hyperactive child."

Worthy's chapter that discusses eye color, sex, and sociability was the stimulus for a chapter that will study behavioral reactivity in relation to sociability and addictive behavior.

A chapter is presented in which eye color and sex are discussed in relation to hyperactivity, to physical/medical conditions in general, to specific learning disabilities, to personality factors, and to developmental skills. Studies are cited in which up to eleven thousand subjects are included from a cross section of economic, racial, and social situations drawn from fourteen southeastern United States school systems.

The final chapter will summarize the findings of the authors and will engage in some speculation concerning eye color, sex, and some general aspects of life that should be of interest to most people. A research design will be suggested for enterprising students of psychology who wish to investigate the eye color phenomenon further, and for those students who maintain that "all the good research has already been done" when it comes time to do their theses and dissertations.

2
Eye color, sex, and kinesics

Bookstores and newstands are loaded with hard-cover and paperback editions containing various treatises on kinesics. The books tell the trusting reader how to improve sexual relations, how to be more success-ful in business, and how to know the innermost secrets of others, all through the study of body language. Most of the books are pure drivel. There are, however, some serious scientists who have spent years studying kines-ics and who have made important contributions to the understanding of this very real phenomenon.

What is kinesics? Most anthropologists would agree that it is the study of body motion as an interdependent part of total communication, and in the context in which it occurs. The scientist who has shed the most light on the subject, both with his own research and through the organization and presentation of the research of others, is Ray L. Birdwhistell. He has brought structure and order to the analysis of behavior in context through the development and use of a highly refined technique of notation used in the description of virtually any body movement. Working closely with distinguished linguists, psychiatrists, and other anthropologists, he has demonstrated with a mass of data that visible and audible behavior are interdependent in the flow of conversations (Birdwhistell, 1970).

In proxemics, which is the study of spatial relations between individuals, various anthropologists have shown that the study of word exchanges *only* is an incomplete measure of social intercourse. But all this does not mean that words are not important. Linguists, psychiatrists, psychologists, and anthropologists who are aware of the cross-discipline work going on in the areas of communication know that everything that happens in context is important to the "meaning of meaning" in human interaction.

The authorities on nonverbal behavior (Birdwhistell, 1970; Hinde, 1972; Jones, 1972; Sheflen & Sheflen, 1972) tend to agree on several things:

(1) that nonverbal behavior is culturally linked both in shape and in meaning;

(2) that no gesture or body motion has been found which has the same social meaning in all societies;

(3) that body motion is a learned form of communication which is patterned within a culture;

(4) that to be able to attach meaning to movement, one must know both the cultural and individual history of reinforcement and the particular context in which the behavior arises.

The latter is a proposition upon which the behavioral psychologists will heartily agree.

In view of the above, it would be absurd to present one isolated research study and to attempt to explain the meaning of nonverbal behavior through the findings. The thesis of the chapter is that eye color, sex, and nonverbal behavior are linked, and that Birdwhistell and others may have overlooked the important factor of eye color in their research.

Sex has been studied across animals and *Homo sapiens* as a "role" factor in nonverbal behavior. The independent variable of sex used in the main study presented in this chapter is not used in that sense, but in a much broader way. Its use in the study accepts the idea that there are both male and female role behaviors that are culturally determined. But it is used also in the belief that at least a part of nonverbal behavior is determined by biological-physiological inherited characteristics of males and females, respectively.

To establish a base or rationale for the original research presented in this chapter, it might be appropriate to review some of the principles of the relationship between eye color and behavioral reactivity and to preview some information that relates both to nonverbal behavior and to other major themes of the book.

Worthy has established that behavioral responsivity is governed by eye color in that the darker the organism's eye, the more reactive it is; or, more accurately stated, the darker the organism's eye, the higher the quality of responses emitted for success at a reactive task. The same is true for light-eyed organisms with regard to self-paced tasks. The preceding statements have to do with what we will call "sending," since this chapter involves communication. There is also an eye color difference in what we will call "receiving"; communication is reciprocal in nature. The important difference is that dark-eyed organisms tend to perceive or to be more sensitive to environmental stimuli than their light-eyed counterparts. The conclusions above are stated in very simple form, but in truth they were reached by a complex investigation involving the various functions of the eye and its relationship to the central nervous system, the endocrine system, and the process of evolution, which will be discussed later in this chapter. For those who wish to pursue the system by which the information was reached, the reader is directed to Worthy's text (1974).

Since Worthy's findings relate to "sending" and "receiving," all of which is not vocal in character, it is believed that his theory is critical to the complete understanding of the nature of nonverbal behavior.

In a later chapter (chap. 5), we will discuss modeling, eye color, and sex. Modeling is a learning process common to virtually all humans in which individuals mimic the behavior of others. This behavior learned by modeling is often not so important in and of itself, although such is the case when musicianship, painting, playing baseball, and so forth, are being modeled. For these are behaviors that are transmitted in complex sets of movements, and modeling is the most efficient method for learning them. For example, it would be very difficult for a person to learn to play baseball well if he had to rely solely on a written or verbal description of how to bat; the behavior needs to be modeled (and also practiced). Inappropriate behaviors are also learned very efficiently by modeling (Bandura, Ross, & Ross, 1963). Sex and eye color are shown in a later chapter to be related to modeling, and inferentially, it is presumed that eye color and sex are related to nonverbal communication by virtue of the fact that attitudes (Bandura et all., 1963) may be modeled nonverbally.

Sociability is also linked to eye color. A doctoral study (Karp, 1972) conducted at Emory University measured the degree to which opinions of light-eyed and dark-eyed male college students were changed by exposure to the opinions of another (female) student. Dark-eyed students were significantly more influenced, as indicated by change in opinion, than were light-eyed students. The dark-eyed students were also more influenced than were the light-eyed students in their liking for the girl by the degree that her opinion agreed with their own. The findings are interpreted to mean that dark-eyed male students are perhaps more responsive to social influence than light-eyed males. Certainly, the higher responsiveness by dark-eyed males to the stimuli in the situation just described is not just confined to the verbal portion of the interaction. In fact, Birdwhistell would probably maintain that a great deal more information of persuasive nature was communicated by nonverbal means.

Sex also seems to be related to nonverbal communication by virtue of being strongly related to behaviorial activity, which, in turn, is related to modeling. To explain the latter connection, dark-eyed persons are more reactive behaviorally. It would follow that such persons would make the best models for the exhibition of specific behaviors one would wish another to replicate.

It also seems to be true from evidence already presented that dark-eyed individuals are more sensitive to environmental stimuli. Therefore, it seems reasonable that dark-eyed individuals would not only make the most efficient models, but would also learn more efficiently than light-eyed individuals by modeling others. This may be true only for largely reactive tasks; however, this point is yet to be thoroughly researched. In a study presented in the chapter on creativity and modeling (chap. 3), the creative task is largely reactive, it seems, and dark-eyed individuals did behave in ways predicted by Worthy's theory. However, it is sometimes difficult to set up a task that is entirely reactive or entirely self-paced. Now to show that sex is related to behavior reactivity and the chain will be complete.

In a study by Jordan (1972), persons were ranked by eye color and sex, two levels on each factor, on how proficient they were on standardized tests of perceptual-motor skills. The perceptual-motor tests consisted of (1) speed and accuracy of finding and marking through specific letters on a printed page, (2) speed of naming 100 color patches, and (3) the Digit Symbol subtest of the Wechsler Adult Intelligence Scale which requires speed and accuracy in using a set of novel symbols. All three of the tasks demand perceptual-motor speed. The test requires high reactive skills as defined by Worthy's operational definition. The proficiency rankings are as follows: (1) dark-eyed females, (2) light-eyed females, (3) dark-eyed males, and (4) light-eyed males. Therefore, sex and eye color interact with the dependent variable of behavioral reactivity. Behavioral reactivity is tied to nonverbal communication because it affects the amount of information one is able both to "send" and "receive." Both sending and receiving are involved with modeling and with learning by modeling; that is, both sides of the modeling situation are involved in some nonverbal communication.

Earlier it was stated that eye color and sex interact. In a statistical sense, "interaction" means that in a multifactor design, such as one that looks at a specific behavior in relation to both eye color and sex, an interaction is said to exist when "as you move from one level to another of a given factor [sex], there is a change [in whatever it is you are measuring] from one level to another on the other factor [eye color]."

It was decided that a test would be designed in hopes of learning more, in an interactional sense, about what sex and eye color have to do with kinesics. The study was designed to show something about "sending" and "receiving," in addition to testing the sex and eye color interaction hypothesis.

Thirty-seven of the thirty-nine students in an under-graduate class in educational psychology were rated on eye color on the first day of class by the instructor. The students were unaware their eye colors were being recorded, and were naive to the experiment throughout its duration. The two students who were not included in the experiment as subjects were used as research assistants. Neither was aware that there was another research assistant in the room. The eye color of the research assistants was also recorded by the instructor.

Subjects were selected for the experiment by the instructor. A random process was used in which the names of the light-eyed males, dark-eyed males, light-eyed females, and dark-eyed females were placed in separate containers. The instructor drew names from each of the containers until the respective cells in the design were filled. (see fig. 2.1)

As the diagram indicates, a three-way fixed effect model, analysis of variance design was used, with the three independent variables being nonverbal communication, sex, and eye color, and with 3, 2, and 2 levels measured, respectively.

Fig. 2.1. Diagram of design

Sex	Eye Color	Kinesics Groups (Nonverbal communication)		
		1	2	3
Male	Dark-eyed	SS 2	SS 2	SS
	Light-eyed	2	2	2
Female	Dark-eyed	2	2	2
	Light-eyed	2	2	2

Factor 1 = Sex
Factor 2 = Eye color
Factor 3 = Treatment dimension (kinesics or nonverbal
          communication)
  SS = Number of subjects

The dependent measures collected were hand gestures, foot movements, and gross body orientation changes: levels 1, 2, and 3 respectively, on the groups dimension.

The observers were instructed separately on how to collect the data; however, they were given no training. They were not told which were the experimental students, but were led to assume that all except themselves were experimental students. Therefore, data were collected on all students. They were not told that a given student's data would only be used from one dimension of kinesic behavior. All three dimensions of behavior were recorded on all students observed.

Part of the course requirement was for each student to select a journal article of his or her choice from a long list of articles, and to abstract the article and make a fifteen-minute presentation of its content and implications to the rest of the class via the lecture method. The instructor took care to see that each student was pleased with the article he or she had chosen, and that the article did not contain material that the student might be uncomfortable in presenting. This was to assure that a fairly typical sample of student behavior would be displayed while presenting the abstract.

It is fairly common knowledge that researchers are able to get highly reliable observations between observers, and this is usually accomplished by highly specific instruction and several training sessions in which observers "practice" recording behaviors and then compare notes prior to making observations in the experiment. The observers in this study were each given the same set of written and verbal instructions on what to record. The hand gestures that were to be recorded were only those that, in the opinion of the observer, were used to clarify or emphasize a point. Foot jiggling or other idiosyncratic foot movements were not to be recorded. Gross body orientation changes (forward, backward, or sideway lean) were the behaviors to be counted on the third level. There was one verbal training session with each observer (held separately) and at the end of the training session they were given a typed sheet of paper with content identical to the verbal instructions. There were no sessions in which they were allowed to practice. The reason for this procedure was to determine if there would be differences in the observers' recordings. The experimenter deliberately selected a light-eyed female and a dark-eyed female as observers. A post hoc comparison was planned between the observations of the two experimenters. The hypothesis was that without practice in observation and recording, a difference in "receiving" by the two experimenters would occur by virtue of their eye colors. It was expected that the dark-eyed observer would record higher frequencies of behaviors on the same subjects than would the lighter eyed observer. This hypothesis was in addition to the hypothesis that there would be more behaviors emitted by the dark-eyed experimental subjects than their light-eyed counterparts, and that this trend would be consistent across the recordings of the two observers.

The observers were asked to rate all dimensions of each student's behavior for two reasons: both students performed in the experiment in lieu of reading and presenting a journal article, and both students were scheduled to do student teaching the following quarter, and the observation and recording experience was believed to be valuable for them in the latter respect.

The seating in the class was a large semicircle. The observers always come to class a few moments early in order to seat themselves at one extreme or the other of the semicircle. The student-observers kept their notebooks tilted slightly away from the student seated either to their left or right to preserve the security of the observation process. Since test items were to be written on the content of the student presentations, all students had notepads and pencils out during the presentations. Thus, it was relatively easy for the observers to remain unobtrusive while making recordings.

The format for each day during the twenty-four-day presentation period consisted of the student's lecture for that day, a discussion of the major topics of the lecture, and suggestions from the class on what content should be included in the test. Since the observations took place over a twenty-four-day span, individual differences in observations brought about by temporary observer conditions were presumed to be largely eliminated. This does not, however, mean that there were not other individual factors affecting one or both of the observers that would account for the differences in observations that were produced.

The data taken by each of the observers were analyzed separately by three-way analysis of variance,

fixed effect model for equal n. Thus, there are two
ANOVA (analysis of variance) tables (see figs. 2.2 and
2.3), one for each of the assistants' observations. Bear in
mind that the data in the two tables represent observations on the *same* subjects. An alpha level of .05 was
chosen for testing.

The three-way ANOVA for observer 1, the dark-
eyed female, yielded significant results for all three main
effects and two of the first-order interactions. The sex by
eye color interaction and the second-order interaction
were not significant.

Fig. 2.2. Summary table of analysis of variance for
observer 1 (dark-eyed female)

Source	df	Sum of Squares	Mean Square	F
Nonverbal behavior	2	1039.00	519.50	58.54*
Sex	1	51.04	51.04	5.75*
Eye Color	1	176.04	176.04	19.84*
Nonverbal behavior x sex	2	102.33	51.16	5.77*
Nonverbal behavior x eye color	2	108.33	54.17	6.10*
Sex x eye color	1	12.04	12.04	1.36
Nonverbal behavior x eye color x sex	2	0.33	0.17	0.02
Residual	12	106.50	8.875	
Total	23	1595.63		

*Exceeds tabled value for significance at .05 level.

Observer 2, the light-eyed female, presented data
which yielded the following results. There were two
significant main effects: eye color and the treatment
factor. Right away it is apparent that the light-eyed
female recorded less than the dark-eyed female. This is
consistent with what one would expect. The same two

first-order interactions and the second-order interactions were significant as was the case with the dark-eyed observer. However, a look at the two sums of squares columns indicate that the cell totals were much greater in the dark-eyed observer's data. This suggests greater sensitivity to environmental stimuli for darker eyed observers alluded to earlier.

Fig. 2.3. Summary table of analysis of variance for observer 2 (light-eyed female)

Source	df	Sum of Squares	Mean Square	F
Nonverbal behavior	2	424.08	212.04	56.54*
Sex	1	16.67	16.67	4.44
Eye color	1	112.67	112.67	30.04*
Nonverbal behavior x sex	2	79.08	39.54	10.54*
Nonverbal behavior x eye color	2	96.08	48.04	12.81*
Sex x eye color	1	13.50	13.50	3.60
Nonverbal behavior x sex x eye color	2	18.75	9.38	2.50
Residual	12	45.00		
Total	23	805.83		

*Exceeds tabled value for significance at .05 level.

So far we have discussed only the differences in the two observers: the "receiving" side of the nonverbal communication that we are interested in. We shall now take a look at the "sending" side of the information transaction. We shall return to "receiving" when the interactions are discussed.

The two analyses of variance revealed significant main effects and interactions. Usually when this is the case, it is a little dangerous to interpret main effect; many researchers are very cagey and omit them altogether and proceed directly to the interactions. In this instance, we will simply say that the three classifications of body movements were significantly different, and that finding came as no surprise. It would be very surprising indeed if an entire group of subjects did not emit more hand gestures than foot movement—or body orientation changes, for that matter—while delivering a fifteen-minute verbal summary of an article. It would not be earth-shaking for one or two persons in either of the groups to display an inordinate amount of one kind of movement as opposed to the others. But for an entire group to emit consistent behaviors across levels would be remarkable. Hand gestures were expected to be consistently higher than the other two categories in frequency, since the subjects were seated and the method

Fig. 2.4. Analysis of variance for observer 1: Graph of first-order interaction—nonverbal behavior x eye color

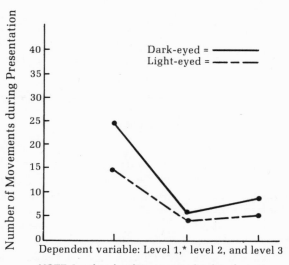

NOTE: Level 1 = hand movements; level 2 = foot movements; and level 3 = body orientation changes.

of delivery was verbal. People tend to use gestures nearest the visual focus of the audience for purposes of elucidation. Hands lend themselves to that condition rather well.

Figures 2.4 and 2.5 present the data of the dark-eyed observer. The first graph shows the amount of "sending" done by dark-eyed and light-eyed subjects, as seen by our dark-eyed observer. As you may notice, the factor of sex is set aside for the purpose of demonstrating the effect of eye color on responses emitted by the subjects across the three body movements in the design. Again, these are as "received" by the dark-eyed observer.

The second graph shows the effect of "sex" on "sending," as seen by the dark-eyed observer. In this instance we have looked again at the same data, but we have omitted the factor of eye color. The graphs demonstrate that both eye color and sex make a difference in the number of responses emitted by our subjects (as seen by

Fig. 2.5 Analysis of variance for observer 1: Graph of first-order interaction—nonverbal behavior x sex

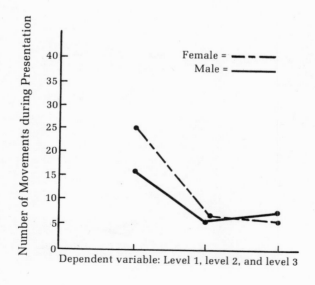

the dark-eyed observer), and especially the number of hand gestures made.

Figures 2.6 and 2.7 show the light-eyed observer's data on the same subjects. The directionality of the lines is similar to that found in the dark-eyed observer's graphs; however, note that the light-eyed observer "received" less from the same subjects than did the dark-eyed observer.

One should *not* be terribly excited about the differences in the observations of the two observers. It is just possible that chance allowed the professor to pick one highly sensitive observer and one who did not see very much. In a situation in which there were large numbers of observers, including the two from this study, the observers chosen by our professor might represent the two extremes of the group in terms of what they "received." However, since the findings are consistent with the behavioral reactivity hypothesis, it might be that experi-

Fig. 2.6. Analysis of variance for observer 2: Graph of first-order interaction—nonverbal behavior x eye color

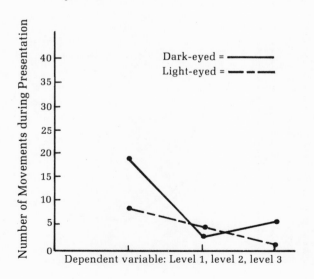

mental observers should be very carefully screened before their data is used in studies such as the one included in this chapter.

In a few words, what do the findings mean? Apparently, the rankings found by Jordan (1972, p. 25) in the study using the test of motor perceptual ability which required highly reactive skills is extendable to body language in the situation just described. Dark-eyed and female persons are more reactive than light-eyed and male; light-eyed and female persons are more reactive than dark-eyed and male persons. Therefore, in studies involving body language, both sex and eye color should be controlled for in both subjects and observers, else the studies might lose some precision. The findings do *not* mean that one can learn, simply by observing the eye color and sex of a person or persons, what will "turn them on," help you attract them, or make a favorable business deal. The research just presented is group research and may not

Fig. 2.7. Analysis of variance for observer 2: Graph of first-order interaction—nonverbal behavior x sex

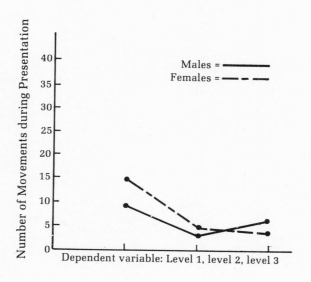

apply at all to a given individual. The appropriate way to treat data of this type is to combine the observations of both observers and perform one three-way analysis of variance. The results of the post hoc comparisons between the observations of the two research assistants are not provided, lest someone take them seriously. Neither were the data combined for analysis. The value of the experiment is that it shows a significant main effect for both sex and eye color in the sensitivity to nonverbal stimuli: an important point indeed. It raises the issue of the effect of training upon "natural" sensitivity to stimuli. It might stimulate a researcher to compare trained and untrained observers while blocking for eye color and sex. Such a design might prove useful and informative. It raises questions about reactivity in general: Are reactive individuals more sensitive to stimuli only; are they motorically more responsive; or are they a combination of the two? The modeling studies in a later section provide some clues.

3
Modeling self-paced and reactive creative behaviors

Creativity has been considered a number of ways, but for purposes of analysis, most creative behavior is usually placed in one of three frameworks: product, process, or personality. In a product analysis of creativity, the *works* of an individual—the painting, the music, the invention, for example—are evaluated by certain criteria to determine their novelty or degree of creativity. This is a direct and a commonsense way of considering creativity, but falls short of our desires as a method. Most psychologists, teachers, and parents would prefer to be able to *predict* an individual's creative potential. In a product analysis, one must wait for the emergence of the product before deciding if it is "creative." A process analysis utilizes what psychologists call "constructs." These are semantic classifications that represent "things" that cannot be directly observed and which probably do not really exist. In addition, they usually have to do with mental functions. A good example would be the "mind." No one has ever seen a "mind," and yet its appearance in our language and writing is ubiquitous. We might say a person has engaged in "divergent thinking" and thus is creative. But chances are very good that we have observed neither the thought nor the divergence. While behaviorists generally do not deny the possibility that such things might exist, they tend to shy away from attempts to measure them due to their unobservability. The third way creativity is often considered is by the personality method. Here, a person's responses to a standard measure of personality or the typical cluster of behaviors that one exhibits under a given set of conditions are evaluated. If they correlate highly with those responses or behaviors by subjects that are already widely accepted as being creative, then a prediction is made based on the degree of correlation. The weakness, of course, is the use of correlation to infer cause and effect. Because two things occur together does not mean one causes the other. This may help us predict with more accuracy but it is not a scientific analysis of behavior etiology. Research in human behavior is fraught with problems such as these.

Until recently, the behavioral school of psychology has avoided the area of creativity. Because of the difficulty in defining and measuring creativity, behavioral researchers have tended to describe all behavior under consideration in the same way, and not to use the term "creative." In this chapter we will present evidence that there is such a thing as "creative" behavior, that creative behavior may be viewed in the two Worthy constructs of self-paced and reactive behaviors, and that operant techniques can manipulate the production levels of creative behavior in a predictable fashion. The thought of manipulating behavior is repugnant to many; the thought of manipulating "creative" behavior is certain to raise the hackles of those in the "freedom and dignity" school or the phenomenological division of psychology. Nevertheless, we shall proceed.

Self-paced creative behavior is operationally defined as that which appears to involve elaborate preparation, great attention to detail, and much rehearsal. Reactive creative behavior is defined as that which proceeds from antithetical situations: that is, it seems spontaneous and is marked by a high degree of novel behavior in response to familiar stimuli, with accompanying humor and high sensitivity to external stimulation. The expectation, of course, is that high self-paced creativity will be associated with light-eyed subjects, and that high reactive creativity will be associated with dark-eyed subjects. Based on previous findings discussed earlier, sex is expected to interact with the treatment groups or conditions under which creative behavior occurs. The operational definition of creativity used in the study is presented later in the chapter and is consistent with definitions that appear in the literature.

Educators have long been concerned with encouraging creative behavior in their students. Unfortunately, very little is known about actual classroom techniques that may be adopted by the teacher to this end. A number of studies have been done on modifying creative behavior; however, none seems relevant to the main points of this chapter: (1) that there are two general types of creative behavior, (2) that these types of creativity are associated with sex and eye color, and (3) that creativity can be brought under the control of experimental conditions.

Behavior modifiers tend to describe behavior in terms of its consequences or types of reinforcement. The assumption is that the antecedents of a given behavior do not control the frequency or intensity of the behavior's recurrence. Instead, what happens just after a behavior determines the nature of the behavior of the organism in the future. The study which is the main focus of this chapter chose three distinct types of reinforcement: noncontingent, individual, and group. Noncontingent reinforcement was chosen as being typical of the traditional classroom setting; that is, creative behavior exhibited by a student is ordinarily not followed by a predetermined and expected set of consequences. Individual reinforcement provides a moderate degree of external control and pressure upon the behavior of concern. It is commonly used in behavior modification studies and is one of the easiest techniques to administer. In such a case, each individual receives a predetermined consequence commensurate with his or her individual performance. Group reinforcement sharing imposes a high degree of external control and pressure. It is commonly used in situations where an individual is emitting highly inappropriate behavior, and is used to bring that person's behavior under control. Ackerman (1972) points out that it is to the group's advantage to help such an individual under reinforcement sharing. Each member of the class gains a reward by helping the student perform in the desired manner. Group reinforcement was chosen to allow the experimenter to investigate the effects of peer pressure and peer approval on creative responding. All three conditions employed lend themselves to replication.

The subjects in the study were eighty-one students from three sections of Adolescent Psychology 3810 at the University of Tennessee. Each student was required to carry out a performance before the class; the performance was to deal with some aspect of adolescent psychology. The students were informed that 20 percent of their course grade depended on the quality of their performances. Each student also prepared a paper for class distribution on the same topic as his presentation. Other options—for example, writing critiques of papers and presentations, reading and critiquing selected writings in adolescent psychology, reading and interviewing over a text, and attendance—were available. Through use of his options, it was possible for each student to attain a grade of A in the course regardless of his score on the presentation. The presentation, however, was required of all students.

Each performance was randomly assigned a presentation date during the quarter. These performances were judged as to their creativity levels by two judges who were randomly selected on a daily basis from the six judges employed. The judges were obtained from another

section of Adolescent Psychology and at no time interacted with the groups in the study. They were not
informed of the experimental conditions in effect in the
various groups they observed. Post hoc questioning of
the judges revealed that they were unaware of the
different treatment conditions in the three groups.
Results of the judges' scoring were available at the end of
each class for the experimental groups only. The judges
were rewarded for their efforts via course credit and
monetary remuneration.

Fig. 3.1. A typical sum-
mary table for
prestudy interrater
reliability estimates

Source	df	MS
Between judges	7	12.85
Within judges	8	1.19
Between scales	1	.07
Residual	7	1.35
Total	15	

NOTE: Reliability =

$$1 - \frac{MS \text{ within judges}}{MS \text{ between judges}} =$$

$$1 - \frac{1.19}{12.85} = .91.$$

Fig. 3.2 A summary of es-
timates of reliabilities
for prestudy

Jan.	15 r = .86	Jan.	31 r = .90
	17 r = .96	Feb.	2 r = .99
	19 r = .95		5 r = .96
	22 r = .90		7 r = .88
	24 r = .91		9 r = .90
	26 r = .83		26 r = .85
	29 r = .88		28 r = .91

NOTE: Four (4) raters per estimate.

The rating instrument (see Appendix) used by the
judges had produced high interrater reliability in previous use. Winer (1971) discusses the use of analysis of
variance to estimate the reliability of measures. His
technique was adopted for use in this study. Winer's
technique applies estimations of reliability across judges
for any given rating (see figs. 3.1 and 3.2). In this manner
problems inherent in this study to test-retest situations
(impossibility of duplicating the presentations and the
shortcomings of videotape; for example, class reactions

are indiscernible) were avoided. Split-half reliability testing was seen as inappropriate due to the variability of behaviors during presentations and was not used. Winer's technique employs a two-way analysis of variance and uses the mean squares from the analysis summary table. Interrater reliability was rechecked five times for each group during the study and found to exceed .82 in each instance (see fig. 3.3). No special training was used for the judges since prior reliability testing had shown no differences between trained and untrained observers (see fig. 3.4). All students in each section were provided a copy of the rating instrument at the beginning of the quarter.

Fig. 3.3. Interrater reliability estimates during
   the study

Group A	Group B	Grooup C
r = .85	r = .82	r = .83
r = .88	r = .85	r = .87
r = .91	r = .83	r = .89
r = .87	r = .92	r = .90
r = .90	r = .95	r = .91

Fig. 3.4. Trained versus
   untrained judges'
   estimates of reliability

Trained Judges	Untrained Judges
r = .83	r = .84
r = .99	r = .99
r = .88	r = .88
r = .85	r = .84
r = .91	r = .90

The Torrance Tests of Thinking Creatively with Words (Form B) was administered to each class prior to the initiation of the experimental condition. Total number of student responses (fluency) was used as the indicator of initial creative levels for each student. Since there was no appropriate way to discriminate qualitatively between responses, these fluency scores were used as the covariate measure in the statistical design.

The instructor's behavior was monitored via a freqency count and held constant between groups and over days. The behaviors monitored were approving and disapproving behavior prior to, during, and following the presentations (see fig. 3.5). The judges were the same as those employed for the creativity scales. Interrater reliability was checked eleven times during the study and found to be 99 percent in each instance (see fig. 3.6). A set method for starting and ending each performance was used. This consisted of giving the performer's name at the start and saying "Let us discuss this" at the end of each performance. The instructor busied himself taking notes throughout the duration of each of the performances.

Fig. 3.5 Instructor's approving and
disapproving behaviors

Perfor-mance	Group A		Group B		Group C	
	Appro-val	Disap-proval	Appro-val	Disap-proval	Appro-val	Disap-proval
1	1	0	1	0	1	0
2	1	0	1	0	1	0
3	1	0	1	0	1	0
4	1	0	1	0	1	0
5	1	0	1	0	1	0
6	1	0	1	0	1	0
7	1	0	1	0	1	0
10	1	0	1	0	1	0
15	1	0	1	0	1	0
19	1	0	1	0	1	0
23	1	0	1	0	1	0

Fig. 3.6. Interrater reliability estimates for
ratings of instructor's behavior

Group A	Group B	Group C
$r = .99$	$r = .99$	$r = .99$
$r = .99$	$r = .99$	$r = .99$
$r = .99$	$r = .99$	$r = .99$
$r = .99$	$r = .99$	$r = .99$
$r = .99$	$r = .99$	$r = .99$

TREATMENTS

Group B was instructed that 20 percent of their total grade would come from the "creative ratings" of their performances. Each student in turn was assigned a creativity (novel behavior) score by the panel of judges.

Each student's grade was based on his own score and those of his classmates. So, all students received the same thirty scores for this part of their grade. The students were informed of this procedure and encouraged to be as creative as possible in their performances. The behaviors specified on the rating scale were made available to all of this group. Reinforcement sharing consisted of the entire group's receiving the daily grade based on the judged novelty level of the individual presentation. Thus, a high score rewarded the entire class, while a low score penalized the entire class. Results were announced in the class period following each presentation.

Group A was identical to group B except for the imposed shared reinforcement. In this case, 20 percent of each student's grade was based *only on the judges' novelty rating of his own performance.* Again, the student received his score during the class period following his presentation.

Group C presentations were rated in the same manner as groups A and B but no reinforcement sharing was imposed and novelty scores were not used in determining grades. The same encouragement to be "creative" was given, however. Grades were assigned on a pass-fail basis, contingent only upon doing the presentation, comprising 20 percent of the student's grade. All assignments of treatments to groups were random.

RATING SCALE

The following criteria developed from Bristol's (1971) article and Torrance's (1971) works formed the rating checklist for the judges.

1. *Effect on the class.*

inattentive   attentive
passive   1..2..3..4..5..6..7..8..9..10   alert

Behavioral criterion—*Alert:* Members of the class had their eyes on the performer, were not engaged in other behaviors, nodded to the performer, made comments, asked questions, took notes, and generally emitted performance oriented behavior. *Attentive:* The students made motions of following and understanding the performance. They sat upright, leaned forward, complied with requests with little time lag, and worked until finished. *Inattentive:* The students did not have their eyes on the performer. They were engaged in other behaviors. They slouched and were extremely relaxed. They may have slept or had their heads on the desks. They made no comments, took no notes, and were non-performance oriented. *Passive:* The students slouched, did not perform to completion, and did not comply with requests.

2. *Appearance of the behavior.* If the behavior had not previously appeared in the class, as opposed to being repetitious of previous performances, the rater recorded a high level of novelty.

has   has never
appeared   appeared
before   ..2..3..4..5..6..7..8..9..10 before

Behavioral criterion—If the performer exhibited a behavior that had not appeared before in this situation, he was being novel. If, for instance, the performer tap danced when no one else had tap danced during their performance, his performance was to be rated as highly novel. The more behaviors that made new appearances, the higher the rating.

3. *Sequencing.* If the sequence of the behaviors exhibited in any performance was expected and predictable as opposed to surprising, unpredictable, or novel, the writer assumed the behavior to be less novel.

redundant, typical,   surprising, unpre-
expected 1..2..3..4..5..6..7..8..9..10 dictable

Behavioral criterion—*Predictable:* The per-

former discussing an exercise after its completion, introducing and labeling a film before
showing, finishing with "Are there any questions?" introducing a speaker and listing his
credentials, explaining the value of part of a
performance immediately on its completion, or
immediately pointing out the relationship
between the performance and the subject matter
was judged as predictable. *Unpredictable:*
Behaviors that were not expected, or typical; for
example, showing two minutes out of a film,
interjecting and unexplained behaviors, indicated unpredictability.

4. *Humor.* Parnes (1967) states that a sense
of humor is a particularly significant trait frequently found in people with creative intelligence. Torrance (1962) states that the humor of a
creative individual may make him even more
unpredictable. Humor was seen as indicative of
novel behavior.

deadpan   1..2..3..4..5..6..7..8..9..10   hilarious

Behavioral criterion—*Humorous:* If the class
was laughing, if people smiled, the performer
smiled, laughed, made jokes, waited for reactions, and so on, he was seen as humorous.
Deadpan: If the performer made no attempt at
humor, did not laugh, did not smile, caused no
laughter or smiles, he was seen as deadpan.

5. *Fluency.* Fluency was the production of
a large number of possibilities or hypotheses
(Torrance, 1971.)

one hypothesis,   ten hypotheses,
one possi-   ten possibility   1..2..3..4..5..6..7..8..9..10   bilities

Behavioral criterion—For any idea or concept,
the score for fluency was directly proportional to
the number of different possibilities or hypotheses. This was rated for each concept the

performer introduced. For instance, if one concept was presented and the performer developed four possibilities for this concept, his rating would have been four. The final score was an average of the total numbers of possibilities for each concept introduced. If the performer, for example, introduced two concepts containing six and four possibilities, respectively, his overall score would have been five.

6. *Flexibility.* Flexibility was seen as the use of many different strategies or approaches (Torrance, 1971).

extremely
inflexible 1..2..3..4..5..6..7..8..9..10   flexible

Behavioral criterion—The number of strategies or approaches used by the performer was directly proportional to his flexibility score. For instance, four strategies would warrant a score of four. These strategies must have shown modal differences. Stating different possibilities or hypotheses in the same mode—for example, oratorically—would be an example of fluency. An example of flexibility would be the use of several different modes of presentation, such as sociodrama, films, tapes, and oration.

7. *Elaboration.* Elaboration was seen as filling out the ideas presented, including details, and making the ideas attractive or embroidering them (Torrance, 1971).

no   ten or more
detail   1..2..3..4..5..6..7..8..9..10   details

Behavioral criterion—The number of details for each concept was directly proportional to the score on elaboration. This was rated for each concept presented. The final rating was an average score.

8. *Preparation.* Torrance (1971) states that a common creative need is "the need to give oneself

completely to a task and to become completely absorbed in it" (p.16). If the presentation gave evidence of being well prepared, well developed, well researched, the performance was judged as more novel.

ill-   well-
prepared 1..2..3..4..5..6..7..8..9..10 prepared

Behavioral criterion—*Preparation:* Films to be shown were present. Speakers were present. Activities were planned. The performer was prepared for questions, manifesting this readiness by immediate responses to questions.

In summary, there were eight judging dimensions of creative behavior. It is fairly easy to analyze each of the dimensions from the self-paced/reactive continuum, and to guess as to which eye color classification would exhibit the most pronounced behavior on each of the scales (more about this later).

RESULTS

Results of the study were analyzed through the use of a one-way analysis of covariance. Homogeneity of regression slopes was tested through the technique described by Winer (1971). The regression slope coefficient, shown in figure 3.7, was not significant ($F = 1.009$, $p = .10$). Hence, regression slopes were homogeneous, assuring that within each group the criterion and covariate measure were related through the same regression constant. The analysis of covariance (see fig. 3.8) showed a significant difference ($F = 183$) between treatment conditions at the .001 level. The Scheffe S Multiple Comparison Test (fig. 3.9) was then applied and significant differences were found between groups A and B F 17.2), groups B and C ($F = 327.6$), and groups A and C ($F = 218.1$).

These results indicate that the treatment groups were significantly different from the control groups with respect to levels of creative responding. The group-shared reinforcement section was significantly superior

Fig. 3.7. Homogeity of regression slopes

Source	df	MS	F
Within cells	75	38.753	
Regression	2	39.104	1.009

Fig. 3.8. Analysis of covariance between groups

Source	df	MS	F
Within cells	77	38.762	
Regression	1	127.282	3.284*
Groups	2	7090.0	183.0**

*$p < .074$
**$p < .001$

Fig. 3.9. Scheffe S multiple comparison tests

Groups A and B
$$F = \frac{[\,(1)\,(55.871) + (-1)\,(62.731)\,]^2}{38.762\,(1/31 + 1/26)}$$
$F = 17.2$, $p < .01$

Groups A and C
$$F = \frac{[\,(1)\,(62.731) + (-1)\,(30.833)\,]^2}{88.762\,(1/31 + 1/24)}$$
$F = 218.1$, $p < .001$

Groups B and C
$$F = \frac{[\,(1)\,(55.871) + (-1)\,(30.833)\,]^2}{38.762\,(1/26 + 1/24)}$$
$F = 327.6$, $p < .001$

to the individual section and both were significantly superior to the control section. The data indicated that both contingent reinforcement groups were under experimental control. The potential range of scores was from eight to eighty. Scores in group A, the individually contingent group, ranged between thirty-nine and sixty-eight. Scores in group B, the reinforcement sharing group, ranged between fifty-eight and sixty-seven. Scores in group C, the noncontingent reinforcement group, ranged between twenty-four and fifty.

ATTITUDES

At the end of the quarter an attitude scale, duplicated in figure 3.10, was administered to all three groups participating in the study. No significant differences were found in the total attitudinal scores of these three groups (see fig. 3.11).

Fig. 3.10 Attitude scale administered to groups A, B, and C

Please respond as frankly as possible concerning the way you have felt about 3810. Place an x at the appropriate place on each line.

1. informative	5..4..3..2..1	uninformative
2. comfortable	5..4..3..2..1	uncomfortable
3. friendly	1..2..3..4..5	unfriendly, hostile
4. free to talk	1..2..3..4..5	not free to talk
5. wanted to talk	1..2..3..4..5	did not want to talk
6. understanding	1..2..3..4..5	did not understand me
7. stimulating	1..2..3..4..5	boring
8. looked forward to coming	1..2..3..4..5	did not look forward to coming
9. got to know many new people	1..2..3..4..5	met no one
10. thought about new, different things	1..2..3..4..5	redundant
11. felt close to group	1..2..3..4..5	did not feel close to group
12. felt supported	1..2..3..4..5	felt threatened
13. wanted to keep on with discussions	1..2..3..4..5	often felt like leaving
14. liked grading system	1..2..3..4..5	hated grading system's guts
15. did not worry about grades	1..2..3..4..5	really worried about grades
16. liked instructor	1..2..3..4..5	hated his lousy guts

Grades in the three sections were similar. Group A, individual reinforcement, had twenty-eight A's, two B's, and one C. Group B, group-shared reinforcement, had twenty-four A's and two B's. Group C, noncontingent reinforcement, had twenty-two A's and two B's.

Fig. 3.11. Analysis of variance: Student attitudes

Source	df	MS	F
Between groups	2	995.04	1.5
Within groups	66	665.01	
Total	68		

NOTE: $\alpha = .05$ $F2, 66 = 3.07$.

CONCLUSIONS

This study demonstrates that a set of responses defined as creative can clearly and effectively be manipulated through contingency management. It also suggests that pressure on the subjects does not have a debilitating effect on creative responses and that grading creative responses may actually enhance the levels of creative responding.

Group-shared reinforcement proved to be the most effective method of managing creative responses. Individual reinforcement, while not as effective as group-shared reinforcement, showed itself to be substantially superior to the noncontingent control condition. Nongraded, low pressure situations in the control group did not enhance creative responding. Therefore, the "freedom of dignity" method of "letting" creativity emerge seems not to be a productive or efficient posture.

SUMMARY

Parnes (1967) summarizes very aptly the current position of that group of researchers engaged in the study of creativity. He and others (Martin, 1971; Torrance, 1963; and Edwards, 1970) assume that there are three definite criteria that situations must meet in order to enhance creative responding. First, the individual must be accepted unconditionally; second, there can be no external evaluation of an individual's product; and third, the individual must receive empathic understanding.

Taking these assumptions in turn, we see that this study did not actively provide unconditional acceptance for any individual. In fact, the reinforcement sharing condition made the individual's acceptability to the group directly contingent on the creativeness of his performance. The performance of this group indicated that the assumption of unconditional acceptance may be inappropriate.

Further, the findings of this study unequivocally demonstrate that external evaluation in the form of grades enhances levels of creative responding.

Thirdly, absolutely no attempt was made at providing any such thing as "empathic understanding," because no feasible or measurable definition could be developed.

The study demonstrated that the treatment conditions did, indeed, raise significantly the number of creative events; the control conditions did not (see figs. 3.12, 3.13, and 3.14). Additionally, the creative responses emitted were judged by criteria found in the literature on creativity. This is important from the standpoint that the experimenter did not set up creativity criteria of his own choosing, criteria that might more easily fit the "eye color" notions of creativity. Instead, he used criteria already established by years of careful research.

Turning to the idea of different types of creativity, is it possible to find differences in responses by eye color, and if so, will the findings be consistent with Worthy's self-paced and reactive behavior constructs?

To make such a determination, the experimental data were reanalyzed by subject eye color. The same procedure used by Worthy and elsewhere in this book was employed to group the subjects into two extreme categories: light-eyed and dark-eyed. In the reanalysis, the experimenter compared the judges' collective dependent measures on each subject by each of the eight

rating dimensions, in turn. Sums and means are shown
for each category, and overall means on the two eye color
groupings are given in figure 3.15. An "eye ball" test
indicates that the overall means are not significantly
different. However, the directionality of higher overall
scores is toward the dark-eyed end. The big differences
in *mode* of responding—whether self-paced or reactive—
are found within the rating dimensions, and this is the
locus of our interest.

Fig. 3.12. Covariate and dependent variable scores:
Group A (individual reinforcement)

Obser-vation	Group	Subject	Covar	Depvar
1	1	1	63	46
2	1	2	66	48
3	1	3	80	62
4	1	4	80	40
5	1	5	53	43
6	1	6	45	62
7	1	7	87	46
8	1	8	83	49
9	1	9	80	62
10	1	10	102	61
11	1	11	117	62
12	1	12	69	60
13	1	13	51	56
14	1	14	118	67
15	1	15	59	59
16	1	16	67	66
17	1	17	46	57
18	1	18	50	59
19	1	19	50	49
20	1	20	64	68
21	1	21	57	57
22	1	22	102	61
23	1	23	50	55
24	1	24	66	54
25	1	25	51	63
26	1	26	65	54
27	1	27	72	67
28	1	28	40	66
29	1	29	61	49
30	1	30	86	39
31	1	31	48	45
		N=31	x̄=69.64	ȳ=55.87

The first scale, *Effect on the class,* is not appropriate to include in the analysis of mode of responding, because it measures the class response (no one individual) and excludes the performer.

The second scale, *Appearance of the behavior,* is the one the raters used to judge the novelty of the performers' behavior. A high score indicates *new* behavior—behavior previously not recorded in class. The mean for the dark-eyed subjects is substantially higher. This seems to be a scale that measures reactive creativity; the behavior is, or has the appearance of being, high in spontaneity.

The third scale is called *Sequencing.* A highly novel

Fig. 3.13. Covariate and dependent variable scores: Group B (group-shared reinforcement)

Obser-vation	Group	Subject	Covar	Depvar
32	2	1	46	62
33	2	2	47	63
34	2	3	60	64
35	2	4	47	61
36	2	5	59	61
37	2	6	54	65
38	2	7	46	65
39	2	8	63	60
40	2	9	76	63
41	2	10	56	60
42	2	11	50	65
43	2	12	121	65
44	2	13	38	58
45	2	14	55	66
46	2	15	67	63
47	2	16	46	62
48	2	17	40	60
49	2	18	71	59
50	2	19	54	66
51	2	20	60	61
52	2	21	49	65
53	2	22	59	61
54	2	23	118	65
55	2	24	52	63
56	2	25	46	61
57	2	26	98	67
		N=26	x=60.69	y=62.73

or unpredictable sequence would earn a high score. Again, the dark-eyed performers' data yielded a substantially higher mean than the light-eyed data. The words "surprising" and "unpredictable" were used as judging criteria; therefore, this scale appears to measure reactive creativity.

Both Parnes and Torrance state that *Humor*, the fourth scale, is a particularly significant trait found in people with creative intelligence. The dark-eyed subjects were again much higher on this dimension. The humor scale, as operationally defined, is clearly one which measures reactive creativity. Very deliberate, deadpan behavior would earn a low score on this measure. An uninhibited, highly spontaneous performance would earn the highest score.

Fig. 3.14. Covariate and dependent variable scores: Group C (noncontingent group)

Obser-vation	Group	Subject	Covar	Depvar
58	3	1	72	33
59	3	2	45	26
60	3	3	128	50
61	3	4	49	32
62	3	5	66	40
63	3	6	79	29
64	3	7	53	29
65	3	8	78	39
66	3	9	35	32
67	3	10	65	29
68	3	11	78	28
69	3	12	75	32
70	3	13	64	39
71	3	14	74	30
72	3	15	70	29
73	3	16	72	25
74	3	17	67	24
75	3	18	57	31
76	3	19	75	24
77	3	20	62	28
78	3	21	86	24
79	3	22	48	31
80	3	23	55	30
81	3	24	45	26
		N=24	x=66.58	y=30.83

Fig. 3.15. Scaled response means by eye color

Categories	Dark-eyed	Light-eyed
1. Effect on the class	$\Sigma_1 = 209$ $\overline{X} = 8.04$	$\Sigma_1 = 252$ $\overline{X} = 8.13$
2. Appearance of the behavior	$\Sigma_2 = 244$ $\overline{X} = 9.38$	$\Sigma_2 = 153$ $\overline{X} = 5.00$
3. Sequencing	$\Sigma_3 = 246$ $\overline{X} = 9.46$	$\Sigma_3 = 146$ $\overline{X} = 4.17$
4. Humor	$\Sigma_4 = 237$ $\overline{X} = 9.11$	$\Sigma_4 = 134$ $\overline{X} = 4.32$
5. Fluency	$\Sigma_5 = 162$ $\overline{X} = 6.23$	$\Sigma_5 = 285$ $\overline{X} = 9.19$
6. Flexibility	$\Sigma_6 = 200$ $\overline{X} = 7.68$	$\Sigma_6 = 236$ $\overline{X} = 7.61$
7. Elaboration	$\Sigma_7 = 153$ $\overline{X} = 5.88$	$\Sigma_7 = 286$ $\overline{X} = 9.22$
8. Preparation	$\Sigma_8 = 128$ $\overline{X} = 4.88$ n (dark-eyed) = 26 $\overline{X} = 60.73$	$\Sigma_8 = 292$ $\overline{X}$ 9.42 n (light-eyed) = 31 $\overline{X} = 57.55$

On the three scales that appear to measure self-paced creativity (Fluency, Elaboration, and Preparation), the light-eyed subjects were clearly superior. In fact, the relationship seems to be ipsative in nature, just as in the competing functions of the eye (resolving power and sensitivity).

The Flexibility scale did not discriminate between dark- and light-eyed subjects. The operational definition requires "different strategies" to earn a high score. This may be the reason it did not discriminate. It seems that one could plan deliberately to have a large number of strategies; it seems just as plausible that one could emit large numbers of spontaneous strategies. In other words, novelty is not a feature of the flexibility scale. As a hint to other researchers, it might be well to further define flexibility so that it can be divided into modes of flexibility.

The study seems to verify that there are at least two different kinds of creativity—reactive and self-paced. Further, the mode of creativity one employs seems consistent with the Worthy eye color theme. It appears also that "good" scales measuring creativity include measures of both kinds of creativity. The scale used by Glover (1973) seems to be such a measure. It also appears that if one is high on one kind of creativity, chances are the same person will be somewhat lower on the other. The most important finding is that the eye color (and mode of responding) of an individual carried over into several aspects of behavior. This is very important in the procedures employed in the analysis of behavior. What it means is that experimenters may be required to block by eye color to add precision to their findings.

A final warning: the reader should not fall into the trap of thinking that dark-eyed people have the market cornered on reactive creativity; nor should they accept the tendency to believe that light-eyed people are the best self-paced creators. The research presented is group research, and is useful only in determining overall trends and making overall predictions. Thank goodness, our population is replete with those who are light-eyed and reactive, and with those who are dark-eyed and self-paced. There are still others who are creative by anybody's measure—self-paced or reactive. Find your mode of creativity and *use it!*

4
Creativity:
one step
further

In the preceding chapter, two probable forms of creative behavior were delineated: self-paced and reactive. Obviously, eye color alone cannot be the single factor we look to in determining whether a person is creative, and, if he or she happens to be creative, what kind of creativity will emerge. In this chapter we shall discuss several factors contributing to creativity, including recent findings on modeling, the effects of praise and punishment, and the role of heredity.

HEREDITY

It would not be terribly surprising to find that there really is some fixed and genetically predetermined upper limit in each person's creative potential. It would be terribly surprising if we should learn enough about it to isolate and quantify it for each individual. Even if we were so advanced that we could now do so, it is not at all clear how this information would help us. Based on what we know about the startling behavioral changes that can be brought about through the manipulation of the environment, we believe that more creative (and intellectual) potential is bound up in our unwillingness to apply techniques of operant conditioning than in our lack of knowledge in measuring IQ. Stated another way, we need not be concerned with the "fixed upper limits" of anybody's intellectual powers until we have exhausted our alternatives in the manipulation of environmental contingencies, which, incidentally, are vast. The one aspect of heredity that we feel is important, eye color and its correlated behavioral "styles," will continue to stimulate our inquiries, but not because we are so naive to believe that the eye color of an individual limits his creative capacities. But rather, we suggest that eye color may largely determine the mode by which creativity is expressed in some individuals. And that is important only in that it may point to ways and means by which the environment may be altered to allow a given individual's most effective response mode to be expressed.

ENVIRONMENT

What do we mean by environment? When we speak of an environment, we are literally speaking of all those stimuli that an organism may perceive. This is an all-inclusive definition subsuming all the physical surroundings, perceptible internal stimuli, other individuals, and the consequences of one's behavior upon the physical, social, interpersonal, and internal environment. In other words, a person's environment is the sum total of his experiences.

Several studies have been conducted investigating the possible effects of manipulation of the environment and consequences of behavior on the level of creative responses a person exhibits. The consensus of these studies seems to be that the environment, which includes consequences of behavior, has a very powerful effect on the level of creativity of people (Goetz & Baer, 1972; Goetz & Salmonson, 1972; Glover, 1973). When we refer to the effect of a person's environment, we refer to the effect of the sum total of one's experiences on one's behavior, and not just those directly affecting the behavior under consideration. With respect to creativity, we believe that we are born with "something" (some predisposition to respond) and that "some things" happen to this predisposition due to the very fact that we live through millions of bits of experience in daily life.

A LEARNED RESPONSE?

Our current position with respect to creativity is that it is a learned set or mode of responding governed only very tenuously by hereditary factors separate from intelligence, and we recognize that the factors that govern intelligence are not well understood. We doubt that there is some "super Q" gene that determines manifest intellect, but we certainly feel that there must be some complex series of gene interactions, or genes that are selected in some combination, that lead to the determination of intellectual abilities; we must admit that we do not know.

Allow us, please, to make the assumption that most of what we do is learned. Further, let us make the operant assumption that we "learn" those responses for which we are reinforced or rewarded, and that we do not "learn" those responses for which we receive either no reinforcement or for which we are punished. Do not ask us to explain how a one-time "successful" response such as the commission of a murder (perish the thought!) is emitted by a person when the person has neither emitted the response before, much less been rewarded for it. Our explanation may not be acceptable to many, but we are of the opinion that modeling has occured either visually, aurally, or verbally (or in some combination thereof) such that an observed response is "stored" for possible future use. This is one of the reasons why we have devoted so much space to the concept of modeling. The notion of modeling and storing responses may play a large part in future research, especially that research focusing on creativity.

Reviewing operant research, we notice that the trend

is always to learn successful or reinforcing responses and to extinguish unsuccessful or nonreinforcing responses. If we examine the traditional classroom and determine successful and unsuccessful behaviors therein, very few behaviors calling for novelty, originality, divergence, or humor (the benchmarks of creative responses) would be adjudged successful. In fact, the most successful behaviors in schools call for conformity, obeisance, and redundance. How often have you been chastised or embarrassed for a "wild" idea in class? How often have you been the chastiser (probably as the result of modeling)? How often have you been told "this is serious business" or something similar and had your speech immediately muzzled? Were you ever punished by "writing off"? Chances are you have, but in case you haven't, writing off is the assigning of one set of behavior (writing) as a punishment for some other behavior. What better way to discourage writing than assigning it as a punisher? How about the assigning of any type of academic behavior as punishment? We can think of no better way to speed extinction of academic and creative behaviors than to punish them or use them as punishers.

Have you ever worked a problem correctly (but with the use of a method not familiar to the teacher) and been told that "this just isn't correct"? And even worse, have you ever had your paintings, essays, poems, or stories laughed at, ridiculed, or misunderstood? If your answer is "yes" to any one of these, we can probably assure you with a great deal of confidence that you are not as creative as you might be. If you feel you are totally lacking in creativity, search your experiences; you may find some very good reasons why you feel this way. It is really no wonder that creativity has been fraught with so many problems for researchers. One of the reasons is that there is such a dearth of it.

Another reason that creativity is so difficult to study is that it is hard to agree on its definition. And at this point, you may well be asking, "Well, what is it?" Creativity has been variously defined by many researchers. Parnes (1967) and Torrance (1963), among others, have tendered definitions. In the previous chapter our system of rating creative behaviors seemed to fit only behaviors that are ongoing in similar situations. We believe that creativity is doing something you have not done before, and something that is useful in some fashion. Torrance (1971) states in a general manner that "to be creative requires learning in addition to recognition, memory and logical reasoning, and such processes as evaluation, divergent production (e.g., fluency, flexibility, originality and elaboration) and redefinition." Hence, if recognition, memory, and logical reasoning are not present, we have an example of noncreative divergence. In other words, any novel response that is not indicative of logic and connected to the problem and its immediate background (recognition and memory) is not creative, but merely novel or divergent. We see this criterion as a goal oriented behavior; that is, creative divergence, as opposed to noncreative divergence, non-goal oriented or unproductive behavior. Doing something that one has never done before may or may not be creative; the creativity comes when the response has been useful. Throwing oneself prone in the middle of the classroom may certainly be novel, especially if it is accompanied by screaming and kicking; but if it is not in some fashion goal oriented and productive, one is merely exhibiting a rather bizarre behavior that is not truly creative. It is true that such behavior may have an instrumental usefulness to the performer (if, for instance, one is relieving tension after Sechenov was misspelled by a student for the 362nd time) and others may not perceive it (the prof has finally gone "bonkers"), but this is often the case with creative works. Who then, and by what criteria, shall decide that a work or a behavior is truly creative?

The final decision on creative behavior really falls into two categories. Works done on a large scale—music, scientific research, sculpture, paintings, and so on—are usually judged by the members of the society in which they occur. For instance, we may feel that we have been extremely creative in writing this book, but the level of creativity herein is not for us to judge; rather, the task falls upon our peers in the psychological community, our general readers, our critics, and (first of all) our publisher.

What about the "little" behaviors that are not aimed at the world at large? With respect to these, we feel that you and perhaps your peers are the only competent judges of whether some behavior has been creative. One may think of a new way, for instance, to make a novel breakfast. If only you consume fried oranges and soft boiled tomatoes, you will have to be the judge of your level of creativity.

How can we honestly, then, make an argument for creativity as a learned behavior? There are a great number of situations that do reinforce creativity. Peers laugh at quips and other forms of humor. Novelty by itself can be extremely self-reinforcing to an individual. How many times have you ever done things differently just to "please yourself" and then felt rewarded for what you did?

Since we are using some terms with which some readers may not be familiar, perhaps we should provide some definitions. Reinforcers or rewards are those stimuli that follow some behaviors and cause those behaviors to appear more frequently under similar circumstances in the future. A reinforcer, then, by definition can literally be any stimulus as long as we perceive it in a "reward fashion." Reinforcers are different for all individuals. What may be reinforcing to you and us may actually be punishing to other individuals. Incidentally, we believe that virtually all individuals have learned the set of creative responding to some degree. We may learn that our puns are only reinforced at parties, during coffee breaks, or at home after the children are asleep. We may have also learned that writing is reinforcing only at certain times, such as writing letters or taking lecture notes. The list of behaviors that can be approached in a creative fashion is infinite, and so are the conditions under which we learn that such creativity is acceptable.

Punishment has the same individualistic characteristic as does reward. No two people experience the same stimulus in the same way. This is true for two primary reasons: one is that our nervous systems apprehend stimuli in varying qualities and quantities; second, since this is true, our "histories of reinforcement" are different—both from the standpoint of how identical stimuli have been perceived and because we have had experiences that are not alike. Masochism is the most concrete example that comes to mind in which some individuals have literally learned that stimuli that cause pain are rewarding. Some individuals, of course, experience the opposite effect. Praise, signs of affection, laughter, and other seemingly universal rewards may actually be very punishing to some people. A little reflection will bear this out. Have you never seen a child in class that seemed to feel uncomfortable at the receipt of the teacher's approval, or, heaven forbid, a kiss from his maiden aunt in front of his peers?

Novelty or creativity, then, seems to appear only in specific situations under specific conditions; that is, organisms learn where, when, and under what conditions to be creative. It may well be that some limits to the level of a person's creative ability are set through heredity, but we firmly believe that creative behavior is learned. Specifically, we believe a predisposition to approach a task in one of two ways is possible: creatively or redundantly. That set of responding or general way of behaving that we call creative seems to be governed by the same rules that govern other behaviors. A creative approach to behavior is taught, punished, rewarded, or ignored, just like any other behavior. We feel, therefore, that creativity can be taught.

EYE COLOR AND CREATIVITY

It was pointed out in the last chapter that there seem to be two discrete modes of creative responding. Reactive creativity is a spontaneous, humorous, flexible, and highly sensitive form of responding. Reactivity is "spur of the moment" and is sensitive to external stimuli that are perceived by the performer. A person who is reactively creative might be best characterized as an adlibber and one who responds very well to external stimuli in a fashion that will maximize the reinforcement value of the response given.

Worthy's (1974) react-approach-flee schema of behavior seems to fit creativity very well. Individuals literally react to their environment, approach the problem in any one of several ways, and flee the situation if necessary. Redd Foxx, Slappy White, and Paul Lynde, to name only a few, are comedians with darkly pigmented eyes. By being able to react to novel stimuli with unusual responses, eliciting new responses from their audiences, and then building more new responses on a rapidly changing pattern of external stimuli, the creativity of these persons is unchallengeable. One of Jonathan Winters's specialties is to walk into a television studio with a wide assortment of props and to become spontaneously a series of characters in response to and with the use of the props. Winters is reactive in that in talk-show conversations he can improvise within one of his standard characters or he can improvise within the character of Jonathan Winters. But Winters is a true genius at comedy and can use both a self-paced and a reactive style apparently with equal comfort, and certainly with equal effectiveness. But more about that later.

It is more communicative to illustrate the types of creativity through the example of comedians, and it is much easier to recognize the exceptions to the light eye-dark eye continuum of reactivity by describing the styles of comics because of their universal appeal and broad visibility. To point out the danger of generalizing to all individuals from group data, perhaps it might be a good idea to mention exceptions, and there certainly are some. One exception to the reactive hypothesis is the comedy of Bill Cosby. Bill Cosby's humor is decidedly calculated, nonrapidfire, and his pace is rather slow. And his effectiveness is not the least bit hampered. This might also be a good place to remind the reader that almost any human endeavor, as it is pointed out in other places in this text, can be accomplished in a number of ways, and that to be either self-paced or reactive does not mean one is either superior or inferior.

Now to come back to the react-approach-flee schema, what about the situation in which a comic is "doing his reactive thing" and it isn't going over so well? It is, we believe, the tendency of the reactive comedian in such circumstances to either flee according to the Worthy paradigm, or shift his responses rapidly in harmony with the perceived shift in external stimuli. Fleeing may not take the literal form of running away in physical fashion; rather, it is the disengagement of an individual from a certain set of responding, removing him from a nonreinforcing situation, or in some fashion attenuating the punishment or lack or reinforcement. Most of us do this by either busying ourselves with some new behavior, pretending that we never made the pun in the first place, or by saying something like "well, I thought it was funny, . . . you had to be there to really appreciate it." Comics sometimes return the punishment to the audience when their responsiveness is not to their liking, using lines such as "Am I going too fast for you?" "Laugh it up, folks, or it gets worse (turning to the wings)." "You didn't tell me this was an embalmers' convention." Then when the audience begins to reinforce the comedian, he shifts back to his material or to his "style" and tries to "punch it up" a little to maintain his hard-won rapport.

But what about the self-paced comic? There are many of them, and, as the late Jack Benny, for example, they usually stay around for a long time. The blue-eyed Mr. Benny was the epitome of self-pacing. Self-paced creativity is well thought out, well planned, deliberate, and designed to meet certain goals. A person who is a self-paced creator is apt to work and rework an item to perfection, relying on preparation and previous experience to maximize the reinforcement potential of his responses. The self-paced creators (comics) are not sensitive to external stimuli to the extent that the reactive performers are during the course of a performance. Instead, their response mode dictates that they withhold the response during the act, but later use the stimuli to rework the material such that it will "do what it's supposed to do." It seems that the self-paced comedian in many ways fits the wait-freeze-stalk tactic

Worthy has ascribed to light eyed organisms. Using Jack Benny and Bob Newhart as examples of self-paced creativity, we see that timing is a crucial element in their comic style. In addition, they both have certain expressions and sight-gags that have worked over and over. With respect to Mr. Benny, the audience knew that whenever Mr. Benny was insulted in some way by one of his stooges, he would place one hand aside his face, say "well!" and do a triple take. Given twenty seconds of silence, Mr. Benny could get five belly laughs with nothing more than "Well!" and facial grimaces and hand gestures. A reactive comedian would have heart seizure while waiting in a self-paced way for twenty seconds to expire. When the self-paced animal is hunting his prey, he waits, freezes, and stalks patiently; he withholds the attack until all the conditions are exactly right. Then he pounces. The self-paced comedian uses the same technique. Mr. Benny "decoyed" with his hip-swinging walk, he gestured, he rolled his eyes and said "Well!"; then with split-second timing, he rendered his audience helpless with laughter. Newhart and Hope use the same deliberate approach and split-second timing. You can almost see Hope mentally "stalking" his audience.

Earlier, the wide range of Jonathan Winters's abilities was mentioned. Winters is dark-eyed, and as such, one would expect him to be reactive. Well, he is; but he is also self-paced. He has prepared monologues and characters that he has "built" over a period of years. In a reactive situation he may go into one of the characters to respond, or he may respond as himself. On the other hand, he uses the characters in a very deliberate and self-paced manner to deliver the comic routine when he is doing a "stand-up." Winters's comic creativity incorporates all the characteristics mentioned in the research literature on creativity, and all the characteristics of both the Worthy response styles. Incidentally, many people do not know it but Jonathan Winters is an excellent painter—a decidedly self-paced form of creativity. In describing Winter's creativity as being at genius level, we were not throwing the term around loosely.

Awaiting, freezing, and stalking, of course, do not occur in creative behaviors in the same sense that they might occur in other overt behaviors. But they are nonetheless apparent. Light-eyed, self-paced indivuals tend to work around definite schedules, waiting for the time when it is necessary for their behavior to start. Self-paced writers, for instance, usually work on rather definite schedules assigning themselves very definite amounts of output for any specific period. Freezing is a behavior that is rather covert for writers, usually occurring as a burst of writing followed by a pause. It is sometimes a characteristic of painters, also. Freezing is usually described in self-report by painters and writers as, "I need to be in the mood." The stalking seems to stem from working at and being reinforced for achieving a particular effect, for example, stalking an audience; or in literal fashion, stalking ideas, concepts, and so on, to be put into words or colors.

MODELING, CREATIVITY, EYE COLOR, AND SEX

As of this writing, we know of only two studies that have investigated the effects of modeling and creativity. Glover (1973) concluded that a model emitting a high rate of creative responding could significantly increase the level of creative responding of the observers.

Gary and Glover (1973) replicated the earlier Glover study and blocked for eye color and sex. The results were rather startling. It seems that the frequency of production (or responses) goes up in positive relation to the frequency of responses modeled. The self-modeling treatment has either a neutral or a negative effect on creative production, depending on the eye color of the modeler. Dark-eyed subjects were markedly negative in their production of creative responses in relation to their baseline responses. Further, the Gary and Glover study showed that the amount of change in levels of responding was very distinct in terms of eye color and sex. Dark-eyed females made the most dramatic change followed by dark-eyed males, light-eyed females, and lastly, light-eyed males. Very evidently, what was being measured in the study was reactive creativity. Dark-eyed individuals responded very strongly to external stimuli, drastically changing their levels of responding in a spontaneous fashion. They directly evidenced external control of their responding (see fig. 4.1). Sex differences also showed a distinct difference in terms of reactive/self-paced behavior, with females being the more reactive (see fig. 4.2). Individuals with light-colored eyes showed a markedly lower level of external control in their behavior.

As of this date no inarguable conclusions can be drawn from this study. It does, however, suggest that eye color is likely to be one of the important variables that should be studied in a number of research projects on the nature of creativity. If future studies confirm the rela-

Fig. 4.1. Groups x eye color interaction

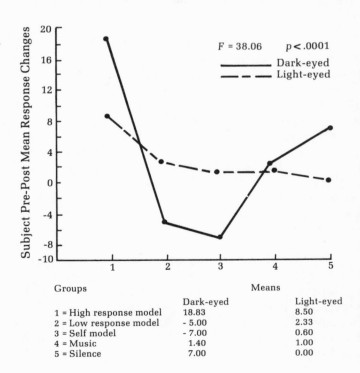

Groups	Means	
	Dark-eyed	Light-eyed
1 = High response model	18.83	8.50
2 = Low response model	- 5.00	2.33
3 = Self model	- 7.00	0.60
4 = Music	1.40	1.00
5 = Silence	7.00	0.00

tionship indicated by the Gary and Glover findings, it would lend strong support to the contention that there are two distinct forms of creative responding, and may point directly to eye color as the major identifier as to which kind of creativity a person is likely to exhibit.

Earlier we dealt with several questions concerning the praise (reinforcement) or disapproval (punishment) of creative behaviors. We concluded from informal observations that creative responding must certainly be a learned response, a response that depends on its consequences in order to determine its probability of reoccurrence. Several studies have appeared that investigate the effects of praise on creative responding. Bohm (1971) examined the effects of nonverbal reinforcement

Fig. 4.2. Groups x sex interaction

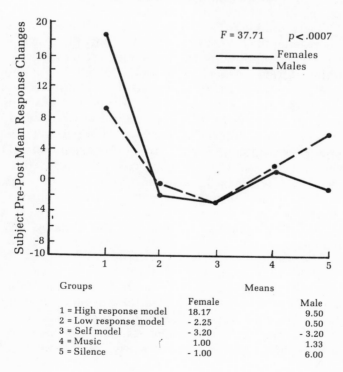

Groups	Means	
	Female	Male
1 = High response model	18.17	9.50
2 = Low response model	- 2.25	0.50
3 = Self model	- 3.20	- 3.20
4 = Music	1.00	1.33
5 = Silence	- 1.00	6.00

of originality scores of disadvantaged first, second, and third grade students. He found a definite increase in originality scores as a result of praise reinforcement. Turknett (1972) looked at the differential effects of an individual reward on the creative productions of children. During the study, children in the individual reward session were told the child thinking of the most unusual ideas would receive one dollar. Tests were scored on the basis of fluency, flexibility, originality, and elaboration. The individual reward condition seemed to "be more motivating," but no significant difference was found. No control groups were used in the study and it was not clear how much of the treatment effects could be attributed to the experimenter's asking for ideas rather than to the experimental contingencies. However, the study does point out the possible effects of group versus individual administration of reinforcers and the possible differences in effect of the two methods.

Goetz and Baer (1972) demonstrated the possibility of increasing the diversity of blockbuilding in preschool children by reinforcing the appearance of new block-building forms for each individual. In the Goetz and Salmonson study (1972), three preschool children were selected on the basis of a lower number of forms used in their paintings. Twenty-five painting forms were identified from other children's paintings. Holding material (paper size, number of paints, size of brush, and so forth) constant, the children were individually exposed to general reinforcement by specific approval of an entire painting; descriptive reinforcement by specific approval for specific forms; and no attention, wherein the teacher watched the painting process without making any comments.

The results demonstrated a definite increase in the numbers of forms used by all three children under the descriptive reinforcement condition. The general reinforcement and no attention conditions seem to have no effect. The conclusion of the study was that it is definitely possible to modify numbers of forms used by students through the use of descriptive reinforcement.

Several other studies have in some fashion applied themselves to the effects of reinforcement on creative responses (Dacey, 1967; Dellas, 1970; Maltzman, Bogartz, & Bregar, 1958; Pryor, Haag, & O'Reilly, 1969; Reese & Parnes, 1970). All these studies demonstrate that reinforcement can be successfully used to raise creativity. None of these studies, however, examined the effects of punishment as a consequence to creative behavior. In a study by Gary and Glover (1974), several possible consequences for creative behavior were administered to a group of 70 undergraduate psychology students at Tennessee State University. The students were randomly assigned to one of seven treatments after having been chosen on a random basis from a population of 190 possible student subjects.

The method consisted of administering an Unusual Uses Test to each student, followed immediately by the treatment and ending with a retest situation. The study has two important factors operating: the use of punishment as one of the consequences to creative production, and the fact that all students (experimental subjects) were rated as being either at 4 or 5 on the 5-point scale of eye darkness. Thus, one of the research questions is Are there differences by treatment conditions between moderately dark-eyed and very dark-eyed subjects? and the other is Does punishment have a differential effect based on eye color? The study controlled for sex, as have many of the other studies, so there were actually four research qestions; that is, one may substitute the word "sex" for eye color in the questions above. It would seem that since no light-eyed students were included in the study the results would describe differences in levels of reactivity rather than on a self-paced/reactive continuum. With respect to the reinforcement conditions, those students evidencing the most change (positive) compared to baseline data ranked themselves in the following way by eye darkness and sex: dark-eyed females, dark-eyed males, moderately dark-eyed females, and moderately dark-eyed males. Thus, the findings are consistent with those of other studies; dark-eyed subjects are more sensitive to external control (reinforcement) than moderately dark-eyed subjects, and females are more sensitive than males. This directional difference is very important since it yields data that are much more discriminative than those in other studies. It also seems to pair extreme reactiveness with extremely dark eyes. At this point we certainly need more evidence before such an interpretation can be accepted as fact. Lastly, punishment: while having a somewhat greater impact on those students with extremely dark eyes, punishment did not, in fact, have a significant influence on the levels of creative responding of any of the students. It seems that eye color does have an important role in the prediction of creative response changes with the use of reinforcement procedures. But earlier we had hypothesized that punishment might lower the levels of creativity appreciably, and the data of this study tentatively negate that stand. It may be that punishment has a differential effect, based on the eye color and sex of the individual. This is certainly an avenue of research that should be explored. Although reward and punishment are known not to have opposite effects it may be that not controlling for sex and eye color has confounded some of the results in previous studies. Based on the assumption that higher external control (reinforcement) is achievable with progressively darker eyed subjects, it seems that punishment would work in the same manner.

In concluding this chapter, let us review some of the ideas we have explored. Part of the review will be rather redundant, but we hope to point out some of the "facts" that might have misled you.

We have presented evidence that eye color and sex both play a large part in the type of creativity seen in individuals. We have shown that creativity is a learned response and as such is subject to the same laws of learning as other behaviors. Generally speaking, darker eyed people tend to be more subject to reinforcement than lighter eyed people; females tend to be more responsive in this respect than males.

Now let us talk about some problems. First of all,

self-paced/reactive research is difficult, the reason being that it is difficult to find "pure" tasks; often an activity will have both self-paced and reactive elements in it. One must guard very carefully against setting up experiments which arrive at false conclusions because of this factor. We have shown that dark-eyed people are more subject to external control than are light-eyed individuals. But we should be very careful how we evaluate those results. We have other evidence which shows that what we are measuring is *immediate responsivity*. We strongly suspect, especially with respect to the modeling procedures of learning, that lighter eyed people indeed do not respond immediately, but that they just might "store" responses for future use. If this is the case, when we obtain a lower measure of creativity under certain reactive conditions for light-eyed persons, we do not know that the lower measure would not be significantly higher if remeasured at a later date. Our hunches about self-paced peoples' behavior and our look at comedians' behavior suggests that an immediate measure of *creativity* under reactive conditions is an unfair and inaccurate evaluation of the light-eyed person's creativity. What we are saying, essentially, is to look carefully at the contingencies of the research design, and be careful about making generalizations that go beyond it.

In this chapter we have tried to present evidence concerning creativity, and have worked from a rather awkward base. First of all, we describe the importance of the environment and we advise not focusing too much on the inherited characteristics of the individual. Then, unhesitatingly, we plunge into discussions of an inherited characteristic: eye color. Most of the people who are keenly interested and actively engaged in the teaching of creativity are the phenomenologists or humanists; yet here we are, behaviorists, talking about how to increase creativity. Phenomenologists and humanists revere the idea of freedom and dignity and the absence of controls. We come on with recommendations about controlling the consequences of creativity as a method of enhancing it.

What is one to believe? It is always our advice to trust the evidence, but not to trust it so explicitly that one stops trying to find new and better ways of "doing the job." It is our belief that the best decisions come from the most accurate and scientific information available. But we are also painfully aware that the best information available is rarely the best information *obtainable*. The message, then, is to test and retest; record data; modify contingencies, test and retest; record data. This is the process by which we refine our decision-making information, and it is the big weakness of the behavioral sciences. Too often decisions are made on the basis of everything other than good information.

5
Eye color, sex, and modeling behavior

In chapter 4, evidence was presented which indicated possible differences between levels of creativity—both self-paced and reactive—that might be predicted by eye color. Three more studies are included in this chapter which provide further evidence that eye color may well be a determining factor in modeling situations. The first study utilizes subjects representing the complete spectrum of eye color. The second research population was made up of students with dark eyes only, but with definite darkness differences. The third is a brief summary of a complex study which strongly supports the notion that eye color is critically related to subject sensitivity to social stimuli in a modeling situation.

Gary and Glover (1974) conducted a simple study dealing with the effect of an audio model on the frequency of appearance of descriptive adjectives. A college classroom of twenty graduate students at Tennessee

State University were rated on eye color on a 1-to-5 scale. The population included eight students with light-colored eyes and twelve with dark eyes. The ratings were collapsed in a manner previously described and in accordance with the Worthy method.

The students were presented a picture of astronaut Neil Armstrong standing on the moon. They were then asked to describe in writing what they saw. Their descriptions were collected and used as baseline data; the dependent measure was the number of adjectives each person used in describing the scene. The experimen-

Fig. 5.1. Modeling change in dark-eyed females

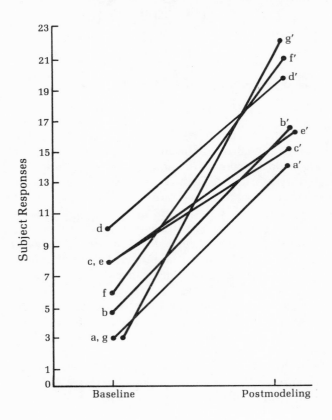

ter then presented an audio tape of a male voice describing the picture. The model used fifty adjectives in describing the same picture they had just seen. They were told, "This the way one of our students described the picture." After hearing the tape, the students were again presented the same stimulus object and asked to describe it again. They were not given specific instructions as to whether they could use the same words the model had used, whether they could repeat words they had used previously, or indeed, what exactly was being measured in the experiment. Except for the stimulus item and the modeling tape, the situation was rather open-ended. There was no reason for them to believe that an

Fig. 5.2. Modeling change in dark-eyed males

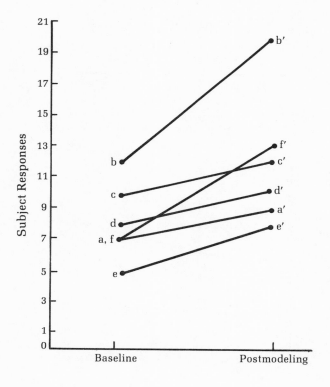

increase in adjectives was expected after the modeling treatment. Graphs depicting their baseline and postmodeling scores, broken down by eye darkness and sex, are presented in figures 5.1, 5.2, 5.3, and 5.4.

Fig. 5.3. Modeling change in light-eyed females

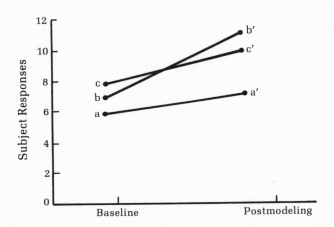

Fig. 5.4. Modeling change in light-eyed males

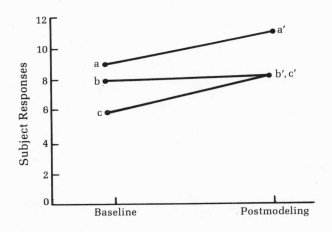

Although modeling is known to have a pronounced effect on behavior, and has heretofore been thought to be rather consistent across subjects, the differences are dramatic when separated in the manner above. The rankings in change are fairly consistent with the rankings by sex and eye darkness found in other studies of human behavior: dark-eyed females, dark-eyed males, light-eyed females, and light-eyed males. Although the sample was taken from a small population, and does not meet stringent requirements of experimental studies, when combined with other findings similar in nature, the results demonstrate the point at hand quite well. External controls do seem to be greater for dark-eyed individuals and females do seem to show greater sensitivity to external stimuli.

What would happen in a modeling situation in which the population was all dark eyed (4s and 5s) but with slight darkness differences within the group? To help answer this question, another study was conducted.

A group of forty-three undergraduate students at Tennessee State University was selected as the research population. They were all Afro-Americans and all were rated as a 4 or a 5 on the 1-to-5 eye darkness scale. The eye color ratings were conducted by impartial judges; interrater reliability was checked throughout the procedure and found to exceed .90 in each instance. The group was then broken into four experimental groups: extremely dark-eyed females, extremely dark-eyed males, moderately dark-eyed females and moderately dark-eyed males. There were thirteen, eleven, ten, and nine subjects in the groups, respectively.

The subjects were then presented an album cover (Santana) and asked to describe what they saw in writing. As was the case in the first study presented in this chapter, the initial responses were used as baseline measures. The balance of the experiment was conducted in the same way. An audio model was presented in which the speaker furnished a description containing fifty adjectives describing the album. The students were then presented the same stimulus object and again asked to describe the album cover. See figures 5.5 and 5.6 for the data and charts of the individual scores.

A simple one-way analysis of variance was used to determine if there were significant differences between the four experimental groups (see fig. 5.7).

The conclusions that are drawn from the findings are as follows: First, dark-eyed individuals seem to be much more receptive to modeling than do light-eyed individuals. Second, females are generally more receptive to modeling than are males of comparable eye color. Third, responsivity to modeling seems to increase as the dark-

Fig. 5.5. Modeling change in extremely dark-eyed females and males

	Sub-ject	Base-line	Treat-ment	Differ-ence
	Females			
	1	4	9	+5
	2	6	10	+4
	3	3	7	+4
	4	2	9	+7
	5	7	11	+4
	6	9	14	+5
	7	2	8	+6
	8	1	7	+7
	9	4	7	+3
	10	3	9	+6
	11	3	8	+5
	12	5	10	+5
	13	4	11	+7
Total	13	53	116	+68
	Males			
	1	3	6	+3
	2	5	7	+2
	3	2	6	+4
	4	6	10	+4
	5	4	11	+7
	6	8	10	+2
	7	6	8	+2
	8	2	7	+5
	9	5	8	+3
	10	4	7	+3
	11	4	6	+2
Total	11	49	86	+37

NOTE: Mean difference: females, 5.23; males, 3.36.

ness of the eye intensifies. Fourth, when the nature of the experimental task is ambiguous, the experimental effect is the same. That is, under conditions in which highly creative and large numbers of responses are solicited for success, the ranking of subject success by eye color and sex is predictable from the Worthy theory. Also, under conditions in which it is *not* known by the subjects that large numbers of creative responses are expected, the modeling effect is the same. This is further support for the "lack of response inhibition" characteristic of dark-eyed organisms, and is supportive of "unintentional" learning via the modeling technique.

Modeling and attention (Glover & Gary, 1974) were applied in an ABA design to examine subject sensitivity to external stimuli. It was hypothesized that there would

Fig. 5.6. Modeling change in moderately dark-eyed females and males

	Sub-ject	Base-line	Treat-ment	Differ-ence
		Females		
	1	3	6	+3
	2	5	7	+2
	3	2	4	+2
	4	7	9	+2
	5	1	5	+4
	6	4	7	+3
	7	4	6	+2
	8	6	8	+2
	9	2	5	+3
	10	2	6	+4
Total	10	36	63	+27
		Males		
	1	5	6	+1
	2	4	6	+2
	3	7	10	+3
	4	2	5	+3
	5	4	6	+2
	6	8	10	+2
	7	5	7	+2
	8	2	4	+2
	9	2	5	+3
Total	9	40	58	+20

NOTE: Mean difference: females, 2.7; males, 2.22.

Fig. 5.7. ANOVA for figures 5.5 and 5.6

Source	df	Sum of Squares	Mean Squares	F
Between groups	3	60.1	20.03	14.3
Within groups	39	54.6	1.4	
Total	42	114.7		

NOTE: .05 3,39 = 4.3126; p .01.

be significant differences in both acquisition and extinction rates of a modeled behavior between dark-eyed and light-eyed subjects. Twenty-three experimental subjects were applied the treatment. The hypotheses were upheld: dark-eyed individuals exhibited steeper acquisition and extinction slopes than light-eyed individuals. Group data were analyzed by regression analysis, and the acquisition and extinction slopes of the eye color extremes were significantly different (acquisition: $P < .05$; extinction: $P < .01$).

The study was complex in design and execution, using six observers: two inside the class and four observing through a one-way mirror. Subject and experimenter relevant behaviors were monitored carefully, with reliability ranging from .73 to .98. It demonstrates the strongest research evidence yet generated by the authors of this book that the modeling effect is significantly more effective for darker eyed than lighter eyed individuals. The most interesting aspect of the study is that although darker eyed persons acquired the behavior faster than the lighter eyed subjects, the darker eyed persons also lost the behavior sooner. This is strong evidence for the hypothesis that darker eyed individuals are more responsive to external stimuli, a postulate important to the linking of reactive human behavior to the react-approach-flee behavior of animals, and one that is critical to establishing that there are phylogenetically programmed differences in human behavior which may be predicted by eye color. The reactive and self-paced question, particularly regarding the human individuals' interaction with the stimulus aspects of the environment, points directly to the notion that eye color is a part of what Skinner calls the contingencies of survival and is not indicative of socially determined, ritualized behavior as many believe.

Eye Color and Responsiveness in the Classroom

After a good deal of thought and speculation about eye color and responsiveness to external stimuli, and after considering the creativity data, it was decided that a look should be taken at what happens to public school students' responsiveness after a number of years in the traditional classroom. Given the predisposition to respond very quickly to external stimuli, do dark-eyed students actually respond more quickly than light-eyed students when confronted with a typical classroom task? Are quick responses in the open classroom rewarded by the typical teacher? If the responses are furnished immediate rewards such as social approval, is this enough to maintain quick responses across, for instance, the elementary grades? Do quick-responding students receive "backup" reinforcers such as good grades at the end of the term—grades which are based on their quick responses in the open classroom? Or do students receive grades that are based on their performances on the deliberate, self-paced tests the teachers devise, and upon their *inappropriate* spontaneous responses in the open classroom? Having been students in a traditional classroom, having observed other students and the nature of their experiences in the traditional setting, and having observed teachers, teachers-in-training, and having conducted teacher training courses, we certainly do not enter into this area as unbiased researchers. In view of all the preceding, it was decided rather than try to answer all these imponderables posed with one complex and clever bit of research, that we would attempt to answer them one at a time. So, backing up to the very first question in this paragraph, a design was hit upon which was believed would give us some information in that regard.

A number of undergraduate teachers-to-be were asked, as part of their observational training, to set a particular condition across a number of classrooms housing fifth, sixth, and seventh grade students. Specifically, they were asked to strike an agreement with the teacher in which the teacher would give the students a bit of new content information taken from any discipline area of the teacher's choice. The new information was to be presented by lecture. Immediately after the lecture, the teacher was to ask a series of ten to fifteen questions covering the main points of the lecture. The student assistant, teacher-to-be, was to tally the number of students by eye color and by sex, who responded to the questions first. No effort was made to score the responses as to correctness. Additionally, no effort was made to see how the teacher responded to the students. The study question was, "Are there differences in speed of response by sex and by eye color to open questions in the classroom?" Before the actual experiment was carried out, the teachers-to-be were asked to rate the eye color of every student in the classroom on the five-point rating scale. They were asked to record how many students there were at each point. By making a seating arrangement chart by eye color and sex, the responses were then relatively easy to mark from almost any vantage point in the class.

The total number of students observed was 2,894. Among those were 1,080 dark-eyed males, 1,044 dark-eyed females, 360 light-eyed males, and 410 light-eyed females. Excluding "ties," "no responses," and other data that could not be used, there were first responses from dark-eyed males 409 times, from dark-eyed females 416 times, from light-eyed males 246 times, and from light-eyed females 238 times. Based on the actual distribution of students by eye color and sex in the experimental population, a theoretical expectation of responses was set up in the four response categories. The differences between actual and theoretical responses were analyzed by the chi-square technique. The result was a resounding 57.26 figure that well exceeds the probability level of

.0001 that the data arranged themselves by chance. The interesting aspect of the finding was that it was just the opposite in directionality to what one might expect from the Worthy hypotheses. Light-eyed males and light-eyed females (in that order) were much more responsive than their proportionate membership would predict. The dark-eyed males and females recorded fewer than their proportionate representation would predict. The finding was somewhat confusing until a great deal of thought was again given to the questions posed at the beginning of this section. In fact, we do not know why the scores arranged themselves thus. We will, however, hazard some guesses, based on reinforcement theory and our subjective knowledge of how the traditional classroom teacher operates.

First of all, it is our guess that the correlation between raising one's hand first in response to an open question and getting called on to answer the question is very low. Next, although dark-eyed persons might initially be more responsive (that is, in the lower grades) and might get called upon to answer questions, it might be that the reinforcing value of teacher responsiveness or approval extinguishes over time because of the absence of backup reinforcement in the form of grades. While we will acknowledge that the teachers' perception of student academic performance is shaped by the students' responsiveness in class, our best guess is that *inappropriate* responses emitted by students are the ones that get remembered and translated into academic ratings on report cards. It is our experience that teachers attend much more to a person's "disruptive" responses in open classrooms than to content or task directed responses. There is still another hunch that might be worth exploring. It has to do with creativity. Our research indicated that the dark-eyed respondents are much more likely to not only be first with a response, but because of the response time element, they are much more likely to be creative in their answers. Such answers are very likely not what the teacher is expecting, and hence, may be "wrong" for the teachers' purposes of evaluation.

There are perhaps many other typical instances in the classroom in which responsiveness is not only not rewarded, but is openly punished. Therefore, the findings from our fifth, sixth, and seventh grade classrooms are not terribly surprising. It would be very interesting to begin at the kindergarten level and proceed directly through the twelfth grade and see at what developmental stage the curve changes. You will note that we are presuming that the child with little or no school experience will respond in accordance with the Worthy hypotheses. A follow-up study that employed essentially the same format as the one just described discovered just that in two church-operated nursery school settings. The only difference between it and the middle school study was that the size of the chi-square was smaller (but still significant) and the directionality was reversed in favor of the eye color prediction. The ranking on quick responses was: dark-eyed females, light-eyed females, dark-eyed males, and light-eyed males. Our best guess is that the onset of puberty marks the change point in the way students are responded to. Puberty marks the point where adolescents are *not* susceptible to familial reinforcement to the degree to which they are susceptible to reinforcement from people and factors in the larger environment. The ability to provide and receive sexual reinforcement, in particular, is important. We invite the reader to investigate this hypothesis; the findings might significantly influence the way teachers respond and grade their students based on competitive responding to the "open classroom" question activity.

The research that has been presented on quick responsiveness, modeling, and creativity seems to set the stage for some far-reaching notions about what should go on in the classroom, perhaps even some moralizing in that regard. Specifically, it seems that we should be able to select some particular teaching strategies based on our student populations' eye colors and upon our instructional goals for them. There is, however, a tremendous amount of research that must be done before we embark with full confidence on such a course.

In spite of the constraints we must impose upon the findings presented thus far, it might be appropriate to elaborate upon the implications of the concepts we have touched upon.

TEACHERS AND MODELING

In the chapter dealing with creativity, some specifics were mentioned that will improve the effect of modeling. One of the unfortunate things about modeling that too infrequently gets discussed is that it is a two-edged sword. Both desirable and undesirable behaviors in a model's repertoire will be emulated. Sometimes teachers have vague and superficial notions about how modeling operates and attempt to set up modeling situations with disastrous results. For example, in the situation in which a student is disruptive or otherwise engaging in unwanted behavior, the teacher punishes the student so that others will "learn what not to do" from the example. The teacher intends for the rest of the students to model negatively; that is, to engage in behaviors opposite to those displayed by the punished student. However well-intentioned the teacher's actions may be, the rest of the students are more likely to model the teacher than the punished student. After all, what kind of a model is a guy who is getting a spanking? Of course, you say, the student will model the person with the power, and you are correct. If the behavior being modeled by the teacher is administering corporal punishment, then the students learn that the use of physical force is appropriate to settle disputes or to control the behavior of others. If the punishing behavior is sarcasm, ridicule, fits of anger, and so on, you may rest assured that these are the behaviors the other students will display at a later date.

Although the main thrust of this text is to present the very interesting phenomenon of eye color and its effect on behavior, we cannot refrain from giving the teacher a little "how-to-do-it" advice from time to time. This is one of those occasions.

We know from findings presented earlier that dark-eyed students as a group are much more responsive to environmental stimuli than light-eyed students. We also know that this translates to mean that perhaps they "see more behavior in others." Not only do the darker eyed students see more, but they tend to react quicker and in a more pronounced way to what they see than their lighter eyed counterparts. We are also very familiar with the fine work of the researchers cited elsewhere in this text with respect to modeling or duplicating behavior. Dark-eyed students then not only make the best models by virtue of the quickness and intensity with which they react, but they tend to learn faster by modeling. The research that has been presented is very clear-cut and dramatic on these points. We therefore recommend that teachers, especially, use this information in the classroom to the students' advantage.

We will not make a specific situational recommendation; but rather, will urge the teacher to think very carefully about the nature of the behavior to be modeled. Next, the teacher should think very carefully about who should be chosen to display the behavior. For example, rather than engage a student to display a behavior (which might come across as artificial), for the benefit of the learner the teacher should try to find someone already displaying the behavior. The teacher should then note very carefully whether the intended learner is a reactive individual or whether the target student is more self-paced. After duly considering the nature of the task, the availability of a model, and the nature of the intended learner (and all in relation to the eye color and modeling data herein), it may be that an appropriate strategy will come to mind. We are not abandoning tried-and-true principles of reinforcement but rather, helping the teacher focus on discrimination stimuli that will produce more effective results.

One of the authors had an interesting experience with respect to the modeling phenomenon. He was engaged as a high school science teacher immediately after his graduation. Heady with his new-found freedom, he grew a handsome mustache (and this was several years before it was fashionable to have "extra" hair on the face). About mid-term a rather harried assistant principal brought it to his attention that fully half the male student body sported similar lip decoration. Several of the parents of these students complained that they were having difficulty inducing lip-shaving behavior in their sons due to the fact that their "teacher has a mustache." The problem eventually resolved itself without the necessity of bringing cold steel to bear on the face of the young science teacher, but this certainly exemplifies the power of a teacher to model behavior and should forewarn the most unobtrusive adult that "somebody's watching you."

EYE COLOR, MODELING, AND THE MEDIA

First, we recommend that all forms of social learning be examined for eye color effect. From both a theoretical and a practical standpoint, this will perhaps lead to a more thorough and precise understanding of behavior. Aside from the eye color and sex categories (with which we have spent so much time) and how they relate to modeling, modeling itself is worthy of discussion.

If various individuals receive varying amounts of learning via modeling, should we not be seriously concerned with the nature of television and radio programming? Indeed, we should. Bandura and Ross (1963) relate the outcomes of a series of modeling studies demonstrating that children definitely produce behaviors that models emit in their presence, and specifically, aggressive and destructive behaviors were modeled very effectively. They also describe the rather startling finding of film-mediated models being as effective as real-life models in transmitting deviant patterns of behavior. Although modeling is more pronounced the more similar to himself the learner perceives the model to be, definite modeling effects have been produced with the use of puppets and cartoons (Bandura & Ross, 1963). And although modeling effects have been detected with the use of punished models (Bandura & Ross, 1962), modeling is more effective if the model is either rewarded or treated neutrally rather than punished.

We could devote a good deal of space dealing with

the specifics of television "entertainment" and commercials, and the ethics of the use of modeling to influence behavior—whether the influence is intentional or unintentional. Efforts have already been exerted in this respect, and some have been successful (the ban of cigarette commercials). However, perhaps our attacks on television have been too zealous and even a little one-sided. We have read very little about the positive effects of modeling on TV, except for a few articles lauding programs like "Sesame Street." No research has been done, and as far as we know, the networks have not been asked to help in conducting research on the possible public benefits of some of the very good morning quiz shows. We refer particularly to the shows which require contestants to furnish quick responses to categories of information. Many of the shows are quite sophisticated, and if we believe the modeling data, we must accept that many of these shows model academic competency.

Especially in view of the eye color and sex data and their relationship to modeling behavior, it would be very interesting to get network executives' responses to entreaties to cooperate with university research teams. We have no reason to believe that just because TV companies are profit motivated they have abandoned all other values for money. It is to the advantage of those people who are charged with the responsibility of picking interesting and responsive contestants to have the best information possible to make those choices. For example, those programs not only require contestants who can furnish correct information to make the contests exciting and competitive, but they also need people who will display highly communicative body language and perhaps throw in an occasional ad lib. Programming would no doubt benefit greatly from the scientific selection of candidates. If TV networks could be encouraged to cooperate with university research teams in the selection of candidates and in the analysis of the responses of contestants, it probably would raise the public image of the TV industry, provide more interesting programming, and might make significant contributions to the science and technology of human behavior. There are literally thousands of research designs that could be of great benefit not only to psychology, but to the industry and consumers of entertainment services. Rather than continue to indict TV, perhaps the scientific community should cooperate with the networks in helping them to exercise their stewardship in productive communication.

Of course, there are many designs that could be carried out without the cooperation, or even the knowledge of, the television industry. For example, the eye color hypotheses could be thoroughly investigated by any researcher willing to spend the time before the tube. But to have the proper experimental controls, and to be able to execute long-term research projects that employ a large number of independent variables, one would almost need to be "on the scene."

The main intent of this section is to try and influence researchers to approach television with something other than criticism. Although the criticisms that have been leveled at TV are certainly well founded, in not having alternatives to offer that have some payoffs for all concerned, we the critics commit the same errors that we have committed in the traditional classroom: We fix on inappropriate behavior, we offer no alternatives, and we refuse to recognize good behavior when we see it. As behavioral scientists we know this does not change behavior and often contributes significantly to its perpetuation. We should also recognize that TV executives are TV executives, and that they know about TV. We are the people who know about behavior and should not expect others to "do our thing" for us.

We have tried to show in this chapter that eye color, an inherited characteristic, and sex designation both have an important effect on learning by modeling. We have also tried to show that this does not negate the tremendously important impact of the environment upon the learner. Neither would we have you believe that what we have shown to be related to sex and eye color holds for each and every single individual; the research presented herein is group research, lest any of you feel uncomfortable. We would prefer to take neither an environmental nor a genetic position exclusively in our explanation of behavior; but rather, we would like to be able to reduce the error term inherent in nomothetic designs. We would like to be able to give practitioners of behavior modification a few more discriminative cues as to "when to do what" and to do it more effectively.

6
Eye color and color versus form perception

Recent eye color and perception studies raise a number of questions of interest to special education. Special education is a division of curriculum and instruction in public schools that deals with the ten basic areas of student exceptionality. The areas or categories vary somewhat from state to state, but in general they encompass the mentally retarded, blind, partially sighted, deaf, hearing impaired, gifted, emotionally disturbed, physically handicapped, multiply handicapped, and those with specific learning disabilities. It is the last category, specific learning disabilities, that may profit from the eye color studies. There are five basic divisions of specific learning disabilities: visual, auditory, visual-perceptual, verbal, and gross and fine motor coordination. Methods and materials, and the organic nature of students who have visual and visual-perceptual problems, seem to be pertinent to the eye color and perceptual studies. The students in these categories have great difficulty in learning to recognize, discriminate, and reproduce letters of the alphabet, numbers, words, and geometric designs. People who are certified to teach such students are highly trained, often holding the master's or doctoral degree with emphasis in the division of learning disabilities. Many will admit, however, that with all the information that has been accumulated over the past few years, the best methods for teaching such children leave much to be desired.

After reading the research of Worthy and the studies done by others that he presents in his text, several questions arise. Do students that have been diagnosed as having visual and visual-perceptual problems exhibit the same reactive and self-paced tendencies in relation to eye color as do "normal" students, and are the competing functions of the eye that are correlated with eye color the same for these special education students as for the general population? If the responses of the special population are consistent in proportion with the general population but just more pronounced, are there variations in instructional methods and materials which can be derived from eye color research that may improve the educational outcomes for special students?

Before presenting the study of the special student population (Gary, 1972), it is necessary to review briefly the rationale for suspecting that eye color is associated with learning problems of special students.

Worthy (1974) describes the visual characteristics of dark-eyed and light-eyed subjects thus: light-eyed people are more sensitive to form; dark-eyed people respond more to color. The competing functions of the eye (resolving power and sensitivity) that are associated with eye color (dark and light eyes, respectively), are also associated with color and form perception skills. Further, the amount of light entering the eye seems to be the determining factor. In his chapter on functions of the eye, Worthy discusses the color spectrum and the length of light rays in relation to eye color for those who wish to read a more technical description of the perceptual differences.

Markle (1972) conducted a study investigating the

color versus form perceptual phenomenon. He used the Rorschach Projective Technique with both black and white and colored protocols. He obtained findings that were consistent with those of Worthy. Dark-eyed subjects tended to respond to colors in the protocols; light-eyed subjects tended to notice the forms.

Fig. 6.1. The Mueller-Lyer illusion

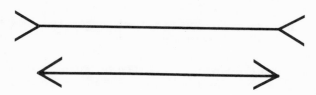

Figure 6.1. is used to test the Mueller-Lyer illusion. Pollack and Silvar (1967) conducted a study in which the protocol was used as the stimulus object. Subjects were asked, "Which of the two middle lines is longer?" The researchers believed that susceptibility to the illusion (the lines are actually identical in length) is related to the subject's eye pigmentation. They studied thirty-five children for whom the amount of eye pigmentation had been determined by an ophthalmoscope. Each child was tested for susceptibility to the illusion. The Pollack findings indicate that lighter eyed persons are more accurate in their perceptions of form and are more sensitive to angle, configuration, and so forth, as indicated by the Worthy hypothesis.

Worthy cites a study by Jahoda (1971) in which dark-eyed subjects were more accurate at reading relief maps in a geography course when colored maps were used.

There is good reason to believe, therefore, that one explanation for visual and visual-perceptual problems of special education children may be associated with the eye color and attendant perceptual characteristics of the children.

Gary (1972) conducted an experiment to determine if teenaged special education students diagnosed as having visual or visual-perceptual difficulties would display the same tendencies with regard to color versus form perception as the "normal" populations in the Pollack and Jahoda studies.

The research population was randomly selected from a large group of students between the ages of sixteen and eighteeen that had all been diagnosed by standardized tests as having visual or visual-perceptual problems. The group was balanced; there were eighteen dark-eyed and eighteen light-eyed students, with nine males and nine females in each eye color grouping. The study was not controlled for race because the Jahoda study was cross-cultural, and did not indicate any differences associated with ethnicity.

Two sets or series of colored 35mm slides were made, with seven slides in each set. The slides depicted geometric figures against a plain white background. The first slide in each series had only one colored geometric figure; the second slide in each series had two colored geometric figures, and so on, up to seven figures in the final slide of each set. In each slide with two or more geometric figures, the figures were unique in configuration, as were the colors. There were instances of repeats in both color and figure from slide to slide, but there were no color or form duplications on a particular slide. The subjects were presented both series of slides for a total of fourteen exposures per subject.

EXPERIMENTAL CONDITION 1: FIXED VIEWING AND SEARCH TIME

In the first experimental condition, the subjects were asked to view the projected slides, in turn, for a specified amount of time. At the end of the viewing time, the subjects were asked to approach a table nearby on which a large number of geometric figures (actually, the same figures used in the slides) were arranged randomly in a display. The students were instructed to select the precise geometric design(s), in form and in color, they had just viewed. They were given a predetermined amount of time, variable with the number of figures on the slide, to make their selections. They were told they would receive a cue from the experimenter when their search time had elapsed, and were encouraged to use all the time allowed for searching.

After the search elapsed, the students were asked to bring their selections to the experimenter for "recording." They were given no feedback on the correctness of their choices. The experimenter recorded the number of form and color errors made by each student on each trial. Seven trials, corresponding to the seven slides, constituted experimental condition 1. Color errors and form errors were the two dependent measures in condition 1.

EXPERIMENTAL CONDITION 2: VARIABLE VIEWING AND SEARCH TIME

Experimental condition 2 was different in that students were not required to view the slides nor to search for the figures for the entire amount of time allowed. The maximum amount of time allowed for each of the activities was identical to that allowed in experimental condition 1. However, the students were given these instructions:

> "If you feel you have viewed the slide long enough to make a correct choice, raise a hand and the experimenter will turn off the projector; you may then go directly to the table to make your selections. You have a certain amount of time to look for the figures. You do not have to use all of the time if you have made your choices. Do not worry about the time; if you use all the allotted time, the experimenter will notify you."

These variations in instructions allowed the experimenter to take two additional dependent measures in experimental condition 2: viewing time and search time.

In both conditions, the order of presentation for the slides was from easy to difficult; that is, the first slide depicted one form and the last slide showed seven forms. Each subject was administered both experimental conditions, with the order of slide set presentation being alternated to counter a "presentation effect."

The data were analyzed by ANOVA, repeated measures technique. The findings were significant beyond the .05 level of probability. In both conditions, the dark-eyed students made more form errors than color errors, and the light-eyed students made more color than form errors. Under experimental condition 2 (variable viewing and search time), the dark-eyed subjects used significantly less viewing and search time than the light-eyed subjects. However, when the performances of dark-eyed students in condition 1 were compared to the dark-eyed performances in condition 2, the dark-eyed students made fewer errors overall when they were allowed to self-determine the amount of viewing and search time. The between-condition analysis of light-eyed students showed no significant differences. They also tended to use all their viewing and search time, regardless of the experimental condition. An interesting observation was made about the students during the fixed time condition. The dark-eyed students often began to fidget toward the end of the viewing time for each slide. Sometimes their eyes would wander from the projection screen. The light-eyed students, however, seemed very attentive to the task stimuli. The dark-eyed students seemd to be anticipating the search or to be eager to engage in the search activity. Dark-eyed students also tended to vocalize while engaging in the search activities. No frequency count was taken, but the darker eyed students seemed to ask for feedback on how well they were doing much more often than the light-eyed students.

Two important points are raised by the study just described. Would early visual and visual-perceptual training of students diagnosed as having difficulties in these areas significantly increase their perceptual competencies if the training took into consideration optimal response modes as predicted by eye color? It would seem to be a worthwhile project for someone to undertake.

Perhaps the training should consist of arranging conditions so that dark-eyed students could achieve success on developmental or readiness tasks with the instructional materials and methods focusing on the use of bright colors and reactive conditions and/or response "speeding." It might be that light-eyed students with such problems should be taught with instructional materials and methods that focus on form and subdued color while emphasizing self-pacing or de-emphasizing on speed. If fairly consistent success experiences could be accumulated in both cases, the experiences might alter significantly the students' feelings toward approaching academic problem-solving situations in later school years. What is being questioned is the present tendency to use a *single* instructional method and highly similar materials, with the choices based more on diagnostic category than idiosyncratic response styles.

The other point raised by the study is that the special education students under consideration do not seem to be "different" in their response styles except in the sense that the characteristics that are correlated with eye color, as in the normal population, are just more pronounced than normal. This suggests that there is a possibility that the conditions of instruction, whether at home or in the very early school years, may be partially— if not almost totally—responsible for the deviations in performances observed. This would not hold, of course, in situations where an organic condition such as a brain tumor or lesion could be located without question. Such situations, however, are rare. Surely behavior modification designs could enhance such students' performance if they were constructed with eye color and behavioral reactivity tendencies in mind. A word of caution is in order here, as it is throughout the text. One should keep in mind that the findings presented and the recommendations made are based on group research. Neither all dark-eyed people nor all light-eyed people possess pronounced reactive and self-paced skills, respectively. The findings are to be considered as typical characteristics of the two groups. Enough diagnostic effort should certainly be exerted with respect to each student before placement in a highly specific program based on the principles of behavioral reactivity and color versus form perception.

RESEARCH DESIGN SUGGESTION

The findings in this chapter suggest that if certain alterations were made in the methods of instruction, the materials used, and the methods of student evaluation, that the educational outcomes would be improved over what is now being produced. But that is yet to be proved. Before instituting the changes suggested herein as a regular practice, it would be very wise to conduct further research to determine if, indeed, the expected changes will take place.

Typical tasks for primary level students that have been diagnosed as having visual and visual-perceptual problems consist of activities designed to teach them to identify, discriminate between, and reproduce letters of the alphabet, numbers, words, and geometric designs. The procedures now used seem to be appropriate in every way except one; and that is, bright colored objects and printed materials are usually the order of the day. The common sense reason for this is the common belief that *everybody* attends to and can learn from brightly colored stimuli. The research evidence of this chapter says that this is likely not true. Perhaps a 3 x 2 factorial design should be executed to answer this fundamental question, and then if the findings are consistent with the research that has been presented, one would feel fairly comfortable with proceeding with a program based on the findings. It is suggested that there be three treatment groups: one group exposed to brightly colored stimuli; another group exposed to stimuli that are black, gray, or very subdued colors; and a third group that is exposed to a mixture of the two stimuli. The other independent variable should, of course, be eye color, and should be blocked on two dimensions: light-eyed and dark-eyed. The dependent variables should be highly specific and measurable educational outcomes. Perhaps a premeasure should be conducted on a number of dimensions for each participating student. After a specified amount of elapsed time during which the experimental treatments are instituted, the postmeasure should be taken on each of the students. The differences in the pre- and postscores on each instructional dimension could then be entered as the dependent variables in the 3 x 2 factorial design. The research findings would then indicate if the principles being transferred from a theoretical to a practical instructional situation hold and are worthwhile to institute as a regular procedure.

The design is a very general one that is applicable to a large number of dependent measures. Just a little reflection on the part of creative special education teachers and supervisors will suggest other and perhaps more relevant research designs.

It is very possible that an early childhood instructional program procedure might be developed with the eye color findings in mind that would lower the symptoms of visual and visual-perceptual problems. In the past, some educators have assumed there was something organically wrong with a child displaying visual or visual-perceptual difficulties. While this may sometimes be true, we are more inclined to attribute perceptual problems to a multicausal explanation. It may be in some cases that pronounced symptoms are indicative of poor instructional procedures and learning histories of punitive consequences to approaching educational tasks. Quite clearly, if students tend to be over- or understimulated by the materials presented in the early years, or if response contingencies inappropriate to the student's response mode are forced upon the student, it is very likely to have a debilitating effect on learning, and perhaps, a complex of symptoms. We also hold that we should be reluctant to proceed with the assumption that there is a central nervous system dysfunction if we do not have explicit and observable medical information to that effect. Therefore, it seems reasonable to pursue a behavioral approach with those children displaying certain learning disability symptoms, especially those who are advanced in their school years, and to use materials, equipment, and procedures with the very young students which are consistent with assisting each child to utilize his or her perceptual strengths. We believe the eye color and color versus form information presented in this chapter worthy of consideration for further research as to its applicability to the practical problems in special educational settings.

7
Eye color, sociability, and addictive behavior

Since eye color appears to influence heavily an individual's responsivity to environmental stimuli, both in terms of what is "seen" and how one "reacts" to what is seen, the question arises as to its effect on interpersonal relations or sociability. This became a particularly pertinent question after examination of the data in chapter 2 on kinesics or body language. Another, and perhaps more serious question, is the one concerning addictive behavior. Addictive behavior has been examined by many researchers and from a number of frameworks. Our best guess at this time is that it is tied to a number of variables. There are many who feel that addiction, particularly to marijuana and "hard" narcotics, is largely a social or socioreligious phenomenon. There has long been a "school" that maintains that the alcoholic is a physical or perhaps a genetic "type." These are interesting hypotheses, but at this point, we consider them to be just that. As behaviorists, basically, we lean toward the explanation of addictive behavior as a learned response. But we have already discussed the caution we feel in not taking into consideration all the possible explanations of behavior. Not to do so, we believe, "puts on the blinders" and virtually assures us that we will not make a significant contribution to a science and technology of behavior if we are only willing to view behavior from one, and only one, perspective. We would like, therefore, to present some admittedly sparse and preliminary data regarding sociability and addictive behavior in relation to eye color.

Sociability

First of all, what do we mean by "sociability"? In general, sociability means the style with which people interact. Do people tend to have a lot of friends, or just a few? Do they attend a lot of parties? Do they enjoy the parties they attend? Do they have many intimate friends, several, or just a few? Are their interpersonal interactions marked by a lot of discord or stress, or are their social relationships rather smooth and stable? These are questions that can only be finally settled by looking closely at a large number of variables, and the reasons "why people are the way they are" are probably almost as many as there are people. But the questions seem to have something to do with activation levels. They seem to be linked to how sensitive one might be to environmental stimuli. They therefore seem to be important to study in relation to sex and eye color.

The first study on sociability involves sex, eye color, and attitude. Milton Rokeach developed a self-report scale that purportedly measures the "authoritarian" personality. The authoritarian personality is the person who tends to be very religious, is likely to take extreme positions on issues, is very sensitive and responsive to authority, among other characteristics. Some of the descriptions of the authoritarian personality, or the clusters of typical behaviors exhibited by such a person, seem to look a great deal like a "reactive" person; that is, highly responsive to external control. With this in mind, a study was designed in which the Rokeach "D" scale (Dogmatism) was administered to a large number of twelfth grade students while controlling for sex and eye color. The groups were: dark-eyed females, light-eyed females, dark-eyed males, and light-eyed males. When the results were analyzed, the findings were not consistent with our expectations. The means of the scores of persons in all but one group were within a few points of each other and were obviously not significantly different. The dark-eyed male group was significantly higher in Dogmatism that the others. It was expected that the dark-eyed females would score highest, dark-eyed males or light-eyed females next, and light-eyed males lowest. Instead, the order from high to low was dark-eyed males, light-eyed males, light-eyed females, and dark-eyed females. Although the last three groups were very close together, the directionality was not at all as was expected. It is also difficult to come up with an explanation why the dark-eyed males were significantly different from the other groups. The difference barely reached the .05 level of probability, and may be explainable for some very good reason not known to the experimenter. In any event, there seems not to be a common-sense way to explain the significant difference obtained in relation to the eye color theory of reactivity. Perhaps the construct "Dogmatism" does not contain a set of specific reactive behaviors, but rather some of both reactive and self-paced.

A questionnaire was then prepared to gather some preliminary data on eye color and social interaction. The questionnaire was administered by a number of university graduate students. Their method was to stop every fifth pedestrian on the street and solicit the information. Each student used a short, standard "spiel" that explained who they were, what they were doing, and made assurances about anonymity of respondents. Of course, in this kind of data collection one cannot claim random sampling, and there were no controls on age, socioeconomic status, and so on. The data are interesting, however, and provide enough preliminary information to suggest that a well-controlled study might be in order. A total of 246 persons responded; 136 were classified as dark-eyed and 110 were light-eyed. Each interviewer asked the person for a self-report on eye color and also rated the person. Other than when there was a discrepancy between the subject's self-report and the interviewer's judgment, there were no reliability checks on eye color. In such an event, the interviewer simply looked at the subject again and made an adjustment in the rating if, in the rater's opinion, it was warranted. The subjects were not controlled for sex.

One of the questionnaire items measuring sociability was:

I enjoy the company of other people (a) always
(b) sometimes (c) never.

Among the dark-eyed subjects the people surveyed answered in the following way: 130 always, 5 sometimes, and 1 never. In contrast, the light-eyed subjects' answers were more evenly distributed among the choices: 14 answered always, 88 sometimes, and 8 never. A chi-square analysis of the results indicates a highly significant difference in the way the scores apportioned themselves. The chi-square ratio was 171.04 ($p < .001$). The directionality of the responses shows there is a pronounced difference in the way dark-eyed and light-eyed people surveyed enjoy interacting with others. The data presented are much too skimpy to formulate such a hypothesis, but it appears that eye darkness may indicate something about stimulus need. Specifically, darker eyed individuals seem to seek stimulation and light-eyed individuals seem to have much lower requirements in this regard.

Another item on the questionnaire was:

I enjoy parties (a) always (b) sometimes
(c) rarely.

The dark-eyed subjects responded with 63, 39, and 34 to the categories, respectively. The light-eyed people answered with 16, 41, and 53. The analysis of the question yielded a highly significant chi-square (28.35 p

<.01). The question seems to be measuring essentially the same phenomenon as the first questionnaire item, but the word "parties" may have altered the distribution of responses somewhat. The differences between the dark-eyed and light-eyed responses seem to be on the first category (always). A much greater proportion of dark-eyed people report they *always* enjoy parties. Although the proportional difference is not as large, a greater number of light-eyed persons report *rarely* enjoying parties. The two groups responded similarly in the *sometimes* category. It appears that there is a greater stimulus need and a greater enjoyment of social stimuli among the darker eyed group. This kind of finding would be consistent with the Worthy hypothesis.

In order to measure the level of intimacy sought among the two eye color groupings, this question was included:

How many intimate friends do you have?
(a) very many (b) a moderate amount (c) few.

The response categories (a through c) were as follows: dark-eyed, 64, 38, 34, light-eyed, 14, 40 and 56. The responses distributed themselves in a way that closely approximates the distribution on the previous "party preference" question. Also, it yielded a significant difference at the same probability level when analyzed with the chi-square technique.

One of the Worthy constructs is the react-approach-flee tendency of dark-eyed organisms. Two questions were included to get a measure of whether human social behavior is marked by any such tendency. Worthy's use of the construct is in reference to survival-type behavior in animals. It seems that if the behavior is naturally selected and genetically determined, it should be manifested in some way in man's social behavior. The item designed to measure the "approach" tendency is as follows:

In making new acquaintances, do you generally make the first move?
(a) yes (b) no.

Among the dark-eyed subjects, 122 said "yes" and 14 said "no"; among the light-eyed respondents, 54 said "yes" and 66 said "no." The tendency is very clear for the dark-eyed subjects to report taking the initiative in exploring the possibilities for social interaction and confirms Worthy's hypothesis.

The question designed to shed some light on the "flee" tendency in social situations is as follows:

Do you usually initiate leaving the
situation if it becomes uncomfortable?
(a) yes (b) no.

Among the dark-eyed respondents, 144 reported that they fled the scene when it became uncomfortable and 22 reported "standing their ground." As in the previous question, the lighter eyed subjects' responses were approximately balanced between "yes" and "no": yes = 59 and no = 51. Again, the Worthy hypothesis seems to be satisfied.

In the event these tentative and preliminary findings are upheld in a more rigorously controlled design, what do they mean? It is our belief that they could significantly influence a number of ideas now held, particularly concerning psychiatric (diagnostic) categories. Psychologists, counselors, and psychiatrists are continually describing the "withdrawn child," the "aggressive child," the "manic" type, and so on. While these categories are perhaps useful in making general kinds of judgments of the state of health of a given individual, the tendency now is to apply one norm or standard in making the judgment. And while this is probably appropriate to do, the judgments rendered might be much more accurate or meaningful if it becomes an established fact that certain people have different response modes or activation levels—modes that are certainly shaped by environmental considerations, but which have a differential base that is established by genetic predisposition. With this consideration in mind, it might become more important to discover what are the most comfortable stimulus conditions for a given individual. Perhaps we can begin to turn more and more of our attention to what B.F. Skinner has suggested, that of arranging the environment in more appropriate ways to the benefit of people, and in consideration of individual differences. Although Skinner would probably be hesitant to include the phrase "individual differences," it seems that we must begin to look more and more for genetic explanations for differences. But at the same time, we must be careful not to fall into the trap of describing differences in terms of superiority and inferiority. Differences do not necessarily mean deficits. Many differences only become deficits if we refuse to manipulate the environmental contingencies.

There are many places we could put the idea of differential response modes to work for us other than in the therapeutic community. Again, what we do with children in the classroom comes to mind. There are so many activities that are school related which involve response modes of teachers and students that it boggles the imagination to think of them. Perhaps the first step is to research thoroughly the react-approach-flee and wait-freeze-stalk concepts to determine if they are in operation in mankind. If the concepts are upheld and are highly correlated with eye pigmentation, then we have a basis for beginning to look systematically for differential ways of contingency arrangement for the maximization of life experiences. There are very great social problems that may be related to not knowing the factors that influence behavior. Marriage is just one example of a long-term and intimate interpersonal interaction, and its success just might be determined by the compatibility of the partners' activation levels. In view of the unhappiness caused by bad marriages and divorce (or bad marriages without divorce for that matter), this seems like a worthwhile avenue for practical research. There are many such areas that are fertile grounds for the researcher interested in practical matters, and we urge the budding researcher to investigate.

ADDICTIVE BEHAVIOR

Does eye color have anything to do with addictive behavior? Perhaps, perhaps not; however, this is another social problem area on which one would certainly be justified in spending some time and effort.

How do people become addicted to tobacco, alcohol, or drugs? We certainly recognize that having the substances in proximity plays a large part in their eventual prolonged usage. But there are many examples of people who have spent large blocks of time in proximity to addictive substances without using them or becoming addicted to them. How do we explain this? Part of the answer is the amount of reinforcement one receives for *not* using them; however, this does not totally explain why many people never use addictive substances. Based on the Worthy hypotheses, it is our belief that there is a predisposition to the use of addictive substances that is predictable by the activation level of the individual. For example, if a given individual is more sensitive to environmental stimuli than another individual, it seems that the social-peer influences would be much stronger than if one were not so sensitive to such stimuli. It seems that if an individual were dark-eyed and highly reactive, that such a person would be much more susceptible to modeling influences than a light-eyed, self-paced individual. Finally, if an individual happened to be dark-eyed and reactive and were exposed to a model that happened to be dark-eyed and reactive also (a good model), it appears that the chances of the "learner" becoming addicted would be much greater. One of the Worthy constructs drawn from his studies of animal behavior postulates that light-eyed individuals are much more capable of response inhibition that those that are dark-eyed. If this is true, when presented with a situation in which the addictive substance is available, and its use is possibly being modeled, then it seems highly likely that the light-eyed individual would be able to inhibit a response to use the substance, while a dark-eyed and highly reactive person might not be able to inhibit the response.

Many researchers believe that the peer reinforcement power is a crucial element in an individual's "choice" to continue the use of an addictive substance. If this is true, eye color as a determining factor goes beyond the first encounter with drugs, for example. Since light-eyed persons are less susceptible to modeling influences and, based on the sociability studies, are less influenced by peer reinforcement by virtue of not desiring a lot of social stimulation, it seems that even in view of the reinforcing value of the substance (physiological) that light-eyed persons would be much less susceptible to addictive behavior. We do not know that this, indeed, is the case. However, it seems plausible, and again, it seems worthy of the expenditure of effort required to make the determination. And again, we urge serious researchers to investigate the problem.

Although it is not the fashion for psychological researchers to talk about their failures, we shall do so. In the questionnaire that dealt with sociability, we attempted to build in some questions that would help us discover some preliminary information on eye color and susceptibility to addictive substances. The items turned out to be so threatening that a large number of the respondents chose not to answer those particular ones. We did get a sizable number of people who were willing to respond, and those responses were consistent with what the eye color hypotheses would predict. However, so many people from the experimental population chose not to answer that we felt the data should not be presented. The questions dealt with whether the respondents actually used an addictive substance, the frequency of the usage, how many times the subject had tried to stop using it, and whether or not the person had been successful in stopping. It was hoped that some connection could be made between eye color (response inhibition) and addiction to the various addictive substances. Again, we would like to suggest that researchers investigate eye color and its relationship to addictive behavior.

Just as a suggestion, it seems that the state departments of vocational rehabilitation might have some fairly good descriptive data ready for processing in this regard. Most state programs are doing work with both alcoholic and drug addicted clients. Some state files may not have client data that provide eye color information; however, most state agencies would be willing to cooperate on a follow-up research project. Vocational rehabilitation personnel seem particularly interested in new and innovative methods of solving the addiction problems. They are particularly interested in methods that will help them pick clients with a high probability of staying off alcohol and/or drugs because this is associated with job success and a closed file. Vocational rehabilitation counselors are evaluated on the number of successful closures they have.

It would also be relatively easy to gather survey data on who has been exposed to addictive substances, and who was either able to refrain from subsequent usage after experimenting once, or who was able to refrain from experimenting at all, even under circumstances where peer pressure was great to do so. Such a study, if controlled by eye color and sex, would give us a good deal of information concerning the part genetics play in the addictive behaviors. It would also be interesting to look at behavior modification programs to determine if eye color is a predictor of success with such techniques. Based on what we know about sensitivity to environmental stimuli, it follows that dark-eyed individuals would be much more controllable via behavior modification than self-paced individuals. Here again, we need hard data to examine, and it seems that researchers in this area should spend some time and effort in this regard.

In this chapter we have tried to touch upon two major areas that are ripe for psychological research: sociability and addictive behavior. They are both areas that are connected with a great deal of human anguish; therefore, they are areas that need no justification for the expenditure of research time, money, and effort. We have suggested that the nature of social interaction styles may be a function of the individual's tendency to be either self-paced or reactive; in other words, to be associated with eye color. We have raised the question as to whether it is proper or correct to view people as falling into diagnostic categories by applying one standard of interaction. We are suggesting that if one has a genetic predisposition to respond to stimuli in a particular way, that it might be more fruitful, scientific, and humane to look more carefully toward manipulating the environment before declaring a person to be afflicted with this or that kind of psychological disturbance.

8
Eye color, sex, and educational diagnostics

I n a study by Dunn and Lupfer (1972), it was shown
that black and white children differ significantly
in reactive skills. To help control for possible culturally
based experiential differences among the students, a new
game was invented by the experimenters from which to
derive the dependent variables. As in the early studies by
Worthy using race as an independent variable, blacks
were found to be superior to whites in the reactive tasks
of the game. In another study involving children (Berry,
1971) eye pigmentation differences were found to be
related to susceptibility to the Mueller-Lyer illusion. The
children in the Berry study were drawn from several
cultures and the findings tend to rule out an experiential
causality to the differences in perception. A study by
Pollack and Silvar (1967) indicates that the susceptibility to the Mueller-Lyer illusion tends to disappear as
children get older. This finding would lead one to believe
that the phenomenon is not only linked to eye color but to

development as well. A study by Gary (1972) presented earlier showed differences in color and form perception that are related to eye pigmentation. In spite of its obvious importance to education and its significance in the quest to determine the influence of genetics and environment on behavior, there is a paucity of research controlled for eye color reported on the behavior of children done across the important developmental years. Certainly, if the behavioral differences that have been found to be related to eye color are genetic in etiology, then they should be found in a pronounced way during the developmental years. In addition, there have been no studies reported in which a large number of subjects has been used. This chapter is devoted to a series of studies in which the N is much larger than one usually finds in psychological experiments, and in which the studies were conducted with school children, many of whom were in the grade range k-2. The grade range k-2 was chosen because it is the time span when very important developmental changes are normally taking place. The developmental changes examined, incidentally, are critically related to subsequent academic performances. In addition to looking at developmental differences in the early grades, the studies focus on physical disabilities and medical conditions, personality factors, and specific learning disabilities, all of which are controlled for eye color and for sex.

HYPERACTIVITY, AUTISM, NEUROTRANSMITTERS, AND EYE COLOR

Before beginning the findings and discussions of the main studies of the chapter, it might be interesting to note some hypotheses raised by Worthy concerning obesity and diabetes. Specifically, Worthy suggests that there may be metabolic characteristics associated with eye color which increase the pobability that certain individuals may have a tendency to become obese or diabetic. Although it is not an established fact that eye color is causally linked to obesity and diabetes, Worthy's speculations concerning the association between eye color and metabolic characteristics seem to have support from the medical profession, and may be of signal importance in devising treatment procedures for the disability of hyperactivity.

Hyperactivity is a condition of children well known to grade school teachers. The hyperactive child is one who seems to attend to all stimuli simultaneously and seems to have no ability to control action impulses. Such a child is unable to sit in a chair for more than a few seconds at a time, is constantly picking at himself, at others, and at objects. Frequently, the behavior goes beyond being disruptive to the organization of the classroom and is rather destructive. Needless to say, such a youngster plays havoc with the organized approach to instructional activities and learning. For many years, the so-called hyperactive youngster was believed to be brain damaged, and there was typically one treatment procedure applied: the administration of various drugs—Ritalin being the most popular among psychiatrists and pediatricians. Later special education teachers discovered that the reduction of light and the amount of stimulation present in the environment lowered the incidence of hyperactivity in such children substantially. The problem was not solved by such procedures for all children, however. Behavioral psychologists took a position strongly against the use of drugs for hyperactive children. The main argument is that the use of drugs brings about changes and side effects that are nonreversible. It has been demonstrated over and over in stringently controlled experimental conditions that behavior modification procedures will reduce the symptoms of hyperactivity for some children with the very important artifact of reversibility—a condition that cannot be met by the treatment procedure involving the administration of drugs. Reversibility means simply that the discontinuation of the use of behavior modification treatment allows the return of the original symptoms. Finally, it became apparent that hyperactivity is multicausal. It was discovered that it is caused in some cases by damage to the central nervous system. Among the specific causes are injury at birth due to delayed or restricted supply of oxygen at parturition, head injury during delivery, usually by forceps, ingestion of lead or lead products at any time, and several other conditions. These all fall under the heading of organic causality. In general, organic causality is nonremediable. If any progress is attained with such children, it usually is a result of programming learning through an undamaged learning modality. The typical approach is simply to try to reduce symptoms through the use of drugs.

It seems now that there may be still another treatment procedure for hyperactivity that is highly effective for certain persons. The treatment procedure suggests that eye color may be an important screening tool in the determination of whether to use a behavior modification approach to the problem or to use the new treatment. Returning to the notion that there is a relationship between eye color, metabolic functioning, and behavior, Dr. Linden Smith, a well known pediatrician, mentioned on his syndicated television show that light-eyed, light-haired, and fair-skinned children seemed to be highly susceptible to a condition called hypoglycemia. Hypoglycemia is a medical condition in which the symptoms resemble a wide range of disabilities ranging from hyperactivity to acute schizophrenic reaction. Hypoglycemia is related to metabolic functioning in that the people who tend to have the disability are unable to maintain the proper blood sugar levels without keeping to a strict high protein, low carbohydrate diet. Typically, the hypoglycemic child (who often is diagnosed as being hyperactive) is overweight and undernourished, is on a high carbohydrate, low protein diet, and is usually light-eyed, light-haired, and fair-complexioned! In his discussion, the pediatrician noted that the manipulation of diet usually caused a disappearance of the symptoms within forty-eight hours, and that the symptoms reappeared with the resumption of the old diet. He further noted that he did not understand why light-eyed, light-haired children seemed to be so susceptible to the hypoglycemic condition. Worthy's findings that eye color is related to metabolic functioning may provide us with insight in regard to this disability. It may be that darker eyed children, by virtue of being more responsive to environmental stimuli, may best be treated by either behavior modification or Ritalin (depending on whether organic damage is present). Eye color may become a "screener" in determining whether to manipulate diet, use behavior modification, or use drugs as the treatment procedure for hyperactivity.

Evidence is mounting that autism, mood disorders, and certain physical disorders including obesity, hypoglycemia and hyperactivity are related much more to the organic *and* inherited state of the organism than to environmental conditions. Happy and Collins (1972) published a study in the *Medical Journal of Australia*. The study investigated the relationship between eye color and the incidence of autism in dark-eyed and light-eyed children. The research was based on two hypotheses:

> (1) a psychological hypothesis which places the autistic child at the extremely introverted end of the behavioral continuum, which ranges from introversion to extroversion;
> (2) a neurological hypothesis linking (a) introversion, (b) the physiology of the ascending reticular activating system (ARAS), especially with regard to its dopamine and neuromelanin content, and (c) the color of the skin, hair, and eyes, through the implication of melanin.

Happy and Collins report biophysical evidence implicating melanin in a protective capacity in nerve cells, and indicate that melanin has a function other than mere pigmentation. There is evidence reported by Happy and Collins that the ARAS uses noradrenaline as a transmitter substance, and that tryosine is a precursor of both noradrenaline and melanin. The researchers argue that a link exists between melanin pigmentation and a postulated defect in the noradrenergic pathways in the ARAS in autistic children. Indeed, their findings were that there is a statistically significant overrepresentation of relatively hypopigmented autistic children and an underrepresentation of relatively hyperpigmented autistic children ($P<.05$). It may well be that some light-eyed people, as a group, have an inherited tendency to autism, and a general predisposition to what Happy and Collins refer to as introversion. This would be consistent with the behavioral construct of self-paced behavior found in light-eyed humans and the wait-freeze-stalk response of lower animals.

In two articles in *Psychology Today*, Brian Weiss (1974) and Jay Weiss, Howard Glazer, and Larissa Pohorecky (1974) discuss the neurotransmitters, diet and hypoglycemia, and various aspects of mood (depression) and behavior (responses to stress). The biochemistry and the behavior discussed in their findings and hypotheses are generally consistent with the Happy and Collins work, and with our speculations about eye color and behavior.

In *The Birth of Language* (Kastein & Trace, 1966) there is a very interesting observation related to eye color in an autistic child. The mother says that when her autistic child was in a very withdrawn state, her eyes which are normally brown became lighter in color, had no depth, lacked luster, and seemed fuzzy. She reports that the child could scarcely learn anything. When the child's eyes became darker in pigmentation and were clear and bright, the youngster would learn rapidly. The mother reports using the cyclic appearance of dark pigmentation in the child's eyes as an indicator, and could, with a fair degree of accuracy, anticipate the amount of contact she would have with the child during a particular period. Between the younster's seventh and eighth years, the cyclical eye pigmentation changes stopped, along with the disappearance of the cyclical learning patterns. Is it possible that the eye pigmentation changes which accompanied the child's responsiveness were related to biochemical changes affective the production of neurotransmitter substances?

THE COOPERATIVE

The studies described in this chapter were made possible through the auspices of an educational cooperative among fourteen school systems. The state in which the studies took place had just passed legislation making it mandatory for all handicapped children, regardless of physical, emotional, sociocultural, or intellectual disability, to be instructed within the regular classroom or, in extreme instances, at least within the school complex. Needless to say, such a program demands radical changes in curricula, administrative procedures, instructional and evaluative activities, and in the roles of teachers. In order to implement properly the new law, it was necessary for the fourteen systems to combine money and efforts. With the enactment of the legislation, the state granted a per-pupil amount of money from minimum foundation funds ("hard money"). However, the amount per pupil was so small and the high cost for the wide range of professional services required just for the identification of the handicapped children made it impossible for the smaller systems to begin to meet the requirements of the law. For instance, one of the small systems received under a thousand dollars as its per-pupil share. Out of this money, if operating in an independent fashion to meet the law's requirements, the school system would have to have hired a clinical psychologist, speech and hearing specialists, provided eye examinations for all children, and purchased college level training for all teachers who were not qualified to teach handicapped students and for all who did not know how to teach by the individualized instruction method. Clearly, that system could not have possibly satisfied the intent of the law. Some of the larger systems could have contracted the needed services, but it still would have been terribly expensive. The school superintendents met and decided to pool their money and, with respect to the law in question, to act as if they were one system. It proved to be an efficient method. The first act of the cooperative was to conduct a survey of all students in the fourteen systems in an effort to locate all the "special" children, screen and diagnose their learning problems, and provide plans for meeting their needs. It was decided that eye color would be recorded for all students in the system, and that control and experimental groups would be picked such that several questions relating to eye color and student characteristics could be asked and answered under controlled conditions.

The total student population of the fourteen school systems is approximately 60,000. As the first step in the implementation of the law, only the grades k-8 were surveyed. There were approximately 43,000 students enrolled in grades k-8, with 11,281 of those students being identified in the first-round survey as "suspected of needing special help." In some of the studies conducted, the entire special population (11,281) was used as the experimental group and the balance was used as the control. In other studies, certain diagnostic categories of the special needs group were used as the experimentals and the controls were randomly drawn from the balance, matched by age, sex, and race.

DEVELOPMENTAL SKILLS

A study was designed in which students in grades k, 1, and 2, and elementary self-contained, special education classes were rated on a standardized observational instrument. The instrument contains a total of seventy rating stems which sample developmental behaviors beginning with age two and continuing through age eight. The developmental areas sampled are (1) communication, with subgroupings of language, differences, number work, and paper and pencil work; (2) socialization (play activities); and (3) motor control, with subgroupings of dexterity (fine finger movements) and agility (gross motor control). The items are stated such that an individual is rated as either "can do" or "cannot do." With the exception of a few items under socialization, all items would clearly be classified as "self-paced" activities under Worthy's operational definitions of behavior. Thus, the instrument allows for the examination of developmental, self-paced behavior in relation to eye color, across a wide range of experiential backgrounds, and on a large number of subjects. The instrument was administered to all students who were tentatively identified as "possibly needing special help," and to a control group as well. All eye colors were represented in both groups. The items were arranged from first to last (within each category and/or subgroup) to be consistent with the way they should appear developmentally. An analysis of the control and experimental results showed the instrument to be sound, developmentally speaking. That is, a positive gradient of means was obtained withing each subcategory in both the experimental and control groups, and the control means were higher than the experimental means in every case, indicating that the experimental group was, indeed, developmentally behind the controls. An intercorrelation table was compiled on the items. The correlations ranged from .59 to .79, demonstrating that if a student is in trouble in one area, chances are high that he or she has difficulties in other areas as well.

Subjects were controlled by sex and by eye color, with eye color categories ranging from 1 to 5 (light to dark, respectively).

A three-way analysis of variance for unequal n was performed on the data. The three independent variables were sex, eye color, and grade level. A separate analysis was done on each of the dependent variables measured by the instrument: language, differences, number work, paper and pencil work, play activities, dexterity, and agility.

The three-way ANOVA yields three main effects, three two-way interaction effects, and one three-way interaction effect. As might be expected, the effect of grade (which looks across grades k-2 and special education classes) showed highly significant differences in each case ($p < .0001$). This finding simply verifies, upon examination of the directionality of the means, that the instrument measures developmental differences; that is, the higher the grade level, the higher the average number of tasks the students were able to perform successfully. The directionality held for the k-2 groups; the special education students consistently fell between the first and second grade performance mean. Again, a verification was obtained: the special education students were appropriately so designated, being considerably below their age norms with respect to the developmental behaviors being tested.

In keeping with the expectations regarding self-paced and reactive behaviors, there were significant main effects on the variable of sex on all the dependent measures ($p < 001$). Females, handicapped or nonhandicapped (grades k-2), were more proficient or scored higher on the developmental tasks than did the males. The exception to this was the male-female difference in the special education classes. The special education male students were consistently better at the tasks than the females. There is good reason to believe that the change in directionality is explainable in terms of sex-role bias rather than some genetic factor. It is very likely that females must display much more pronounced symptoms before being evaluated and placed in a special education class. One of the factors in "getting evaluated" has to do with symptom visibility and, specifically, the degree to which a student disrupts the class. The typical sex-role of the female in the classroom dictates a shy, retiring behavioral stance. Hence, female students present fewer symptoms and thus are less likely to be troublesome to the teacher. Being troublesome to the teacher raises the likelihood that one will be recommended for evaluation. It is reasonable, therefore, that only the most handicapped females would be recommended for evaluation and placed, and would therefore raise the female symptom norm above the male. This is not to say that the bias extends in only one direction. Indeed, males need only display a minimum of symptoms to be evaluated and placed. The male bias is in the opposite direction, thereby enhancing the possibility that a sex difference in developmental skills would occur in a special population.

There also were significant main effects by eye color

on four of the seven developmental dependent measures. The findings bear out the Worthy hypothesis in that the categories in which the stems measure behaviors that are all self-paced in nature (language, differences, number work, paper and pencil work), the lighter eyed subjects, both control and experimental, were at a higher developmental level than the darker eyed subjects. For this analysis, eye color was collapsed into three levels: blue, green, and brown. When the means were arranged, there was a clear gradient or trend from dark to light in the degree of development present. Again, the special education students were different, but by sex category only. The developmental trend with regard to eye color was the same as with the k-2 students. This indicates that the sex and eye color factors are independent variables. In three of the analyses, eye color and sex interacted significantly with the dependent variables.

In summation with respect to the developmental measure, the findings are as follows:

(1) The measure seems to be valid in that it measures developmental differences. Support for this conclusion is found in the fact that a progressive gradient is present across grade levels, and that the special education students scored at a developmental level consistent with their standardized achievement grade level placement.

(2) The instrument measured self-paced skills, and the progressive gradient that appeared by eye color is consistent with what the Worthy hypothesis would predict: light-eyed students develop faster at self-paced tasks than dark-eyed students.

It might be interesting to note that the developmental levels of all k-2 students, regardless of eye color, stabilize at the same level at the terminal point of second grade. Apparently, the instrument measures factors that all students learn, but at different rates. This lends strong support to the proposition that there are genetic causalities for differing developmental levels and for the

Worthy thesis in general. The findings should lead to a modification concerning expectations for the appearance of developmental skills, with the critical factor being eye color. The educational implications are that instructional activities should be keyed to the norm developmental rate associated with the students' eye color.

Apparently, eye color is the best and most reliable predictor of melanin implication in the ARAS (Happy & Collins, 1972). Therefore, it may be possible very soon to begin to make discriminations in treatment procedures for various mood and behavioral disorders, and to gain access to some of what Skinner calls the contingencies of survival, by making simple observations of eye color.

Because the study on developmental behaviors involved such a large n, and because it included all students in the grade levels tested, including those who scored abnormally low, it was decided that the study would be repeated, but with "normal" subjects only.

Seventy-two normally scoring subjects were randomly selected by a computer technique, twenty-four at each grade level k-2. There were twelve males and twelve females selected at each grade level, with six light-eyed subjects and six dark-eyed subjects comprising each group of twelve.

The data from the original developmental skills measuring device was used. There are seven scales on the instrument: language, differences, number work, paper and pencil work, play skills, fine dexterity, and gross motor coordination. Each of the seven categories has ten behaviorally stated stems, worded such that the rater is able to score the child on each stem as to whether he or she "can do" or "cannot do" the task. Each child has a total score within each of the seven categories; the dependent measure is the number of tasks in that category the child can perform successfully.

Subjects were blocked by eye color and sex on two levels and by grade level (k-2). The analyses were seven different three-way ANOVAS, fixed effect model for equal n.

Fig. 8.1. Developmental skills means

		Lang.	Diff.	N.W.	P.P.	Play	Dext.	Agil.
Dark-eyed	Male	2.944	3.166	3.277	4.44	4.66	5.388	2.88
	Female	2.888	3.499	3.388	4.22	4.77	4.611	2.66
Light-eyed	Male	2.833	3.44	3.333	4.111	4.277	4.833	3.22
	Female	4.22	4.27	3.888	3.111	2.44	2.777	2.66

It was hypothesized that the play and dexterity categories (mainly "reactive" stems) would produce higher scores for the darker eyed subjects, and that the balance of the categories (mainly "self-paced" stems) would yield higher scores for the lighter eyed subjects. The dark-eyed and light-eyed means are shown in figure 8.1.

There were significant differences in the appropriate direction ($P < .05$) by grade level in each of the seven ANOVAS, indicating that the instrument is valid. On the main effect of eye color, there were significant differences ($P < .05$) in the areas of language, differences, paper and pencil work, play and dexterity. Light-eyed subjects were better on the first three categories: language, differences, and number work, but with the last difference not significant. Dark-eyed subjects were better at play, dexterity, and paper and pencil work, but with the last difference not significant. Light-eyed subjects were better at gross motor agility, but not significantly so. The hypotheses were upheld except in the instance of paper and pencil skills. Dark-eyed subjects were better in this area. A review of the stems indicates the nature of the tasks is quite mixed in that area, as is the case in gross motor agility. This probably accounts for the lack of significance in both sets of means.

The study does provide a fair measure of inherited behaviors (developmental skills) that are acquired by normally developed persons. It shows significant differences in skill levels by eye color during crucial developmental years, and adds considerable power to the notion that there is a behavioral predisposition toward either self-paced or reactive skills, depending on subject eye color.

LEARNING DISABILITIES

Specific learning disabilities is an area of student exceptionality that poses the greatest instructional problems for parents, teachers, and students. It is particularly frustrating for teachers because many of the disabilities are not detected during the students developmental years when they are most amenable to remediation. Quite often, it is only after students have gone through several unsuccessful school years that a learning disabilities problem is diagnosed. When such is the case, there is usually a firm expectation of student failure on the part of the teacher, the student, and the parent, which makes it difficult for the student to reverse the pattern. Even when conditions of instruction are ideal, results from special instruction are often most disappointing. The learning disabled child's achievement is spotty, and the techniques that work with one child on a specific task will not work with another. This makes it very difficult to plan a program for a learning disabled student. The learning disabled probably constitute the largest exceptional group of students needing special help.

The educational cooperative in which the main studies of this chapter took place identified 9,564 students who were suspected of being in one or more learning disability categories. The instrument used for the first round assessment was a checklist covering the five major areas of learning disabilities: visual, auditory, visual-perceptual, and gross and fine coordination. Only two of the categories, visual and visual-perceptual, are included in the analysis of eye color. The visual category measures pure visual reception or resolving power, and the visual-perceptual category measures both resolving power and discrimination. Resolving power or pure visual reception and visual discrimination are the two competing functions of the eye that Worthy maintains are associated with eye color; specifically, light-eyed persons are allegedly better at discrimination and dark-eyed persons are said to be better at resolving power or pure visual reception. It was believed, therefore, that investigation of these two categories of disabilities might shed some light on the incidence of learning disabilities with respect to visual functions, and in relation to student eye color.

The 9,564 students who were suspected of being in one or more categories of learning disabilities were sur-

veyed by the instrument, along with 305 control students. The control students were selected by the computer, and were drawn on the basis of matching age, sex,
race, and eye color by a proportional formula which took
into consideration the relative proportion of subjects in
the experimental group on the factors listed above. The
total N consisted of 5,552 males and 4,012 females. Eye
color representations were distributed thus: light blue or
gray = 1,339; blue = 2,423; green = 2,279; brown = 2,128; and
dark brown = 1,395. Since the nature of the disability determined the persons who would be the experimental students, their eye colors were also predetermined. The control students were then matched proportionately. The eye
color proportions represented in the study are indicative
of an important characteristic of the learning disabled.
The normal expectation for eye color distribution is that
substantially more than half of the people one might pick
randomly from the street would have dark eyes. In the eye
color distribution cited above for the learning disability
study, it may be readily seen that the eye color distribution is not as we might expect. There are fewer dark-eyed
students than we might expect, and there are more light-
eyed subjects. A reasonable preliminary hypothesis
might be that light-eyed students are more likely and
dark-eyed students are less likely to be learning disabled. In a very general sense this does not support the
Worthy hypothesis concerning the relationship between
eye colors and specifically, the relative efficiency of certain eye colors on the competing functions of the eye. It is
believed that most of the learning disabilities involving
vision have to do with discrimination rather than with resolving power. Light-eyed persons are supposed to be
better at discrimination tasks than dark-eyed persons;
conversely, dark-eyed individuals are supposed to be
more efficient at resolution than light-eyed persons. The
theoretical expectation, then, would be that most of the
students selected out as learning disabled in the visual
modality (excepting pure visual reception) would be
dark-eyed. This did not occur. A disproportionate number of light-eyed students were selected as being dis-

Fig. 8.2. Learning Disabilities

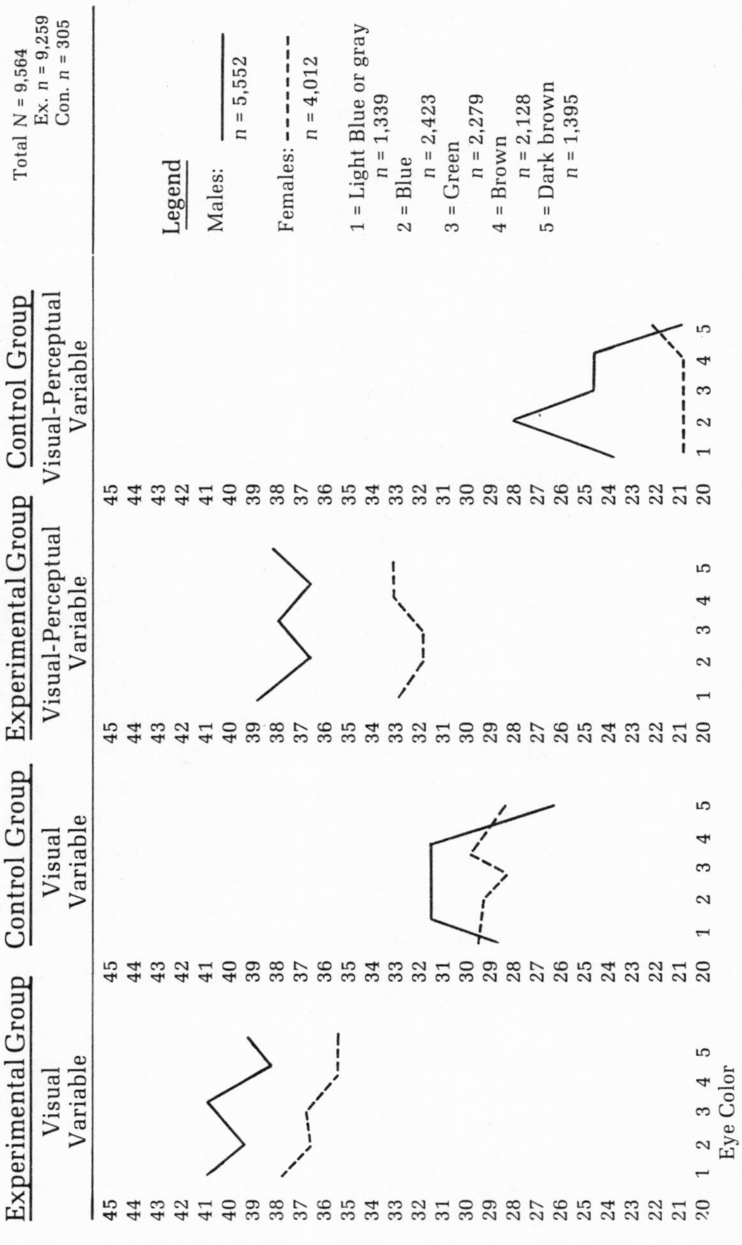
Experimental Group
Visual Variable
Control Group
Visual Variable
Experimental Group
Visual-Perceptual Variable
Control Group
Visual-Perceptual Variable
45 44 43 42 41 40 39 38 37 36 35 34 33 32 31 30 29 28 27 26 25 24 23 22 21 20
1 2 3 4 5
Eye Color
Total N = 9,564
Ex. n = 9,259
Con. n = 305
Legend
Males: n = 5,552
Females: n = 4,012
1 = Light Blue or gray n = 1,339
2 = Blue n = 2,423
3 = Green n = 2,279
4 = Brown n = 2,128
5 = Dark brown n = 1,395

abled, and, as is evident in figure 8.2, they scored on test items as being more disabled than dark-eyed students.

In addition to demonstrating that light-eyed students are not as a general rule better than dark-eyed students at discrimination tasks, it shows a trend that would lead one to believe that dark-eyed students are better at both of the so-called competing functions of the eye (as compared to light-eyed students). This does not mean that the proposition that there are competing functions of the eye does not hold; rather, it seems to show that the levels of effectiveness of the two functions are different depending on eye color. The ipsative nature of resolving power and discrimination may still hold; but, comparatively speaking, darker eyed persons seem to be superior at both, regardless of whether they are "disabled" or "normal."

Very importantly, the results show a significant difference $(P < .001)$ between the experimental and the control groups. This demonstrates that the instrument is reliable in distinguishing between those that are disabled and those that are not. Subsequent standardized testing of the students who were designated by the checklist as being learning disabled validates the checklist findings.

There were also significant main effect findings by sex category in both the experimental and control groups, and on both the visual and the visual-perceptual dimensions $(P < .001)$. Females are consistently better than males at visual and visual-perceptual functions with only one exception: extremely dark-eyed males in the control group on the visual-perceptual variable scored better than the females of that group.

The only first-order interactions were between the dependent variables and sex and the dependent variables and eye color, indicating that the factors of sex and of eye color are independent. There were no three-way interactions.

The conclusions of the study are these:

(1) dark-eyed students seem to be less likely to be learning disabled in the visual modality, regardless of sex

(2) contrary to what has been heretofore accepted, light-eyed student not only are more likely to be learning disabled in the visual modality, but they are not better at discrimination tasks *or* at resolving tasks

(3) rather than just being superior at pure visual reception, it seems that dark-eyed students are better at *both* resolving power and discrimination

(4) males are much more likely than females to be learning disabled, and females are also superior to males with respect to both resolving power and discrimination in the "normal" population.

PERSONALITY FACTORS

As a part of the assessment procedure, a personality scale was constructed consisting of five subcategories. Each category is made up of ten stems which are behaviorally stated and on which each student was rated by his or her teacher. The five categories are: gifted/creative, withdrawn, aggressive, leadership behaviors, and behaviors associated with mental disturbance. The personality measure was administered to an experimental and a control group to determine if the measure actually distinguished between exceptional children and "normal" children, and to determine if there were differences predictable by eye color and by sex category. If the Worthy hypothesis is valid concerning the reactive behaviors, there should be a strong positive association between being aggressive and having dark eyes, and there should be a negative relationship between being withdrawn and being dark-eyed. Since there are at least two types of giftedness and creativity—self-paced and reactive—it was not presumed that eye color would predict giftedness and creativity (the gifted/creative subcategory contains both self-paced and reactive behaviors). It was suspected that there would be no significant difference by eye color on the leadership subcategory.

Finally, it was not presumed that eye color would predict emotionally disturbed behaviors (the stems contained both reactive and self-paced type behaviors). As was suspected, eye color did not predict any of the personality types. There were significant differences on each of the dimensions by sex between the experimental and control groups. Again, the differences, it is believed, may be explained in terms of sex-role rather than some genetic factor. There are some stems in the aggressive subcategory that very clearly are self-paced; therefore, those stems that measure reactive responses were analyzed by eye color separately. There were significant differences between the experimentals and the controls, and there were significant differences between eye color groups 1, 2, and 3, and eye color groups 4 and 5. The group containing eye colors 4 and 5 were significantly more aggressive in both the experimental and control groups than the eye color group containing persons rated as 1, 2, and 3. In this instance, the Worthy hypothesis was upheld: aggression and behavioral reactivity are closely associated; aggressiveness as a personality trait and behavioral reactivity are predictable by eye color. The reactive stems in the leadership, emotionally disturbed, and creative/gifted subcategories were selected out and analyzed by eye color. No significant differences were obtained. The self-paced stems were also selected out and reanalyzed in all subcategories by eye color; likewise, no significant differences were obtained.

The personality factors section of the instrument does not seem to be particularly useful in the prediction of personality type by eye color. It is reasonable to assume, and is perhaps accurate, that personality factors are much more dependent upon one's history of reinforcement and its interaction with genetic and sex-role determinants, and that it is not a fruitful activity to try to predict personality type by eye color alone. The few reactive-aggressive stems that do predict, however, are important in that they verify the notion that the dark-eyed are more responsive to external stimuli and are less likely to inhibit responses.

PHYSICAL DISABILITIES

Prior to discussing findings on eye color and physical disabilities across the ten-county area via the assessment instrument, it might add considerable credence to the idea that eye color and physical disorders may be linked. A study by Bart and Schnall (1973) investigated the relationship between eye color and types of carcinomas and malignant melanomas (cancers). The researchers claim that light eyes are uncommon in patients with solidly pigmented black or dark brown basal-cell carcinomas. All eye colors are found in patients with malignant melanomas. And further, that a solidly pigmented black or dark brown malignant lesion, in a patient with blue, gray, or green eyes, is more likely to be a melanoma than a basal-cell carcinoma. On one level, the importance of the study results suggests that eye color may be an important clinical tool in establishing the type of cancer one is observing. Making the distinction is no mere academic exercise, for the surgical techniques for the two types are quite different, with one being radical in nature. On another level, the study provides more evidence on the relationship of eye color to another human condition, and suggests there may be a strong inherited link in the incidence of the disease.

One of the sections of the cooperative assessment instrument that yielded the most interesting finding is the one dealing with the incidence, by sex and eye color, of diagnosed medical/physical conditions. The results on this section are so tilted in one direction that the experimenter was prompted to go back to the data and run it through the computer again. Essentially, the medical section of the assessment form listed a number of areas in which one might have a diagnosed medical condition: cardiovascular disorder, endocrine disorder, central nervous system disorder, for example. Included in the medical assessment section is a portion devoted to physical anomalies that would not necessarily be considered medical "conditions," but are nonetheless abnormal. Examples are: unusually obese, thin, short, tall, premature or delayed appearance of secondary sex characteristics, and so forth. Dividing eye color ratings into light (1, 2, and 3) and dark (4 and 5), the ratio for incidence of diagnosed medical conditions and incidence of physical anomalies is about 30 to 1, with the light-eyed group being much more susceptible to the physical disabilities and abnormal physical/developmental characteristics. This finding is certainly one which should be checked in other populations. Since the research population in the study reported herein cuts across rural and urban areas, and includes a wide economic range for both black and white racial groups, it is not believed that the medical conditions reported are a function of economics alone. The nonmedical observations of physical anomalies seem to support this belief. The finding is so unusual (or is so at variance with one's expectations) that it would warrant replication in some other geographic area. If the findings hold up in other areas, there might be lines of inquiry established which would be of great significance to future medical research.

The findings of the studies conducted by the Southeast Tennessee Educational Cooperative are extremely important. The studies demonstrate quite conclusively that the sex factor is very important to consider when carrying out educational diagnostics for the purpose of constructing individual educational programming.

The somewhat startling finding that light-eyed students are much more variable in their physical characteristics and in their susceptibility to medical conditions and to learning disabilities is of signal significance.

The relationship between eye color and metabolic functioning and the relationship of both to the susceptibility of light-eyed students to the medical condition of hypoglycemia are also extremely important. It paves the way for a multitreatment approach to hyperactivity, a condition that has been known for some time to be multicausal in nature. Just as important, it provides a rationale for the reduction, if not abolishment, of the use of drugs in the treatment of hyperactive children except in those cases in which a central nervous system lesion can be located.

The fact that developmental skills that are eventually learned by all children (except the more severely retarded) are learned at different rates, and that these rates are predictable by eye color, should affect substantially the way individual educational programming is established.

Finally, such a large study (involving large numbers of subjects) yields firm evidence that both sex category and eye color must be taken into consideration in all studies of human behavior in the future, lest we continue to publish results that have these factors confounded in the treatment variable or hidden in the error term.

9
Speculations

Although the bulk of this text is "data based," we have engaged in some speculation from time to time. Our speculations, though, have been clearly identified as such, and have proceeded directly from our findings or those of others. In this the final chapter, our speculations will have fewer hard facts upon which to rest. There are several reasons for the inclusion of a chapter filled with speculations. First, we wish to demonstrate how infrequently a factor such as eye color is identified which lends itself to so many areas of human inquiry. It has been a long time since a factor has been located that appears to be related to so many aspects of behavior: response speed, visual-perceptual skills, color and form perception, body language, and modeling, to name just a few. Second, eye color can provide research on individual differences a new start, and we can move away from the worn-out battle of the IQ. With the Worthy frame of reference, we can begin to carry out new research designs, we can talk in other than hushed tones about inherited differences, and all without having to rank people. There is already enough accumulated evidence to demonstrate that many of the unidimensional evaluations of persons' intellectual and behavioral capacities do not take into consideration the various ways of responding, and hence, the various modes of success. We have also shown that even though there seem to be wide inherited differences in response skills, environmental manipulation has a vast potential for modifying behavior. Finally, we wish to contribute to the potential for optimal student growth through principles of behaviorism and the psychology of individual differences.

INDIVIDUALIZED INSTRUCTION

The main thrust of the text has been that there are individual differences that are apparently genetic in origin. Further, these differences seem to "make a difference" in the way people see and respond to the world around them. We have tried to convey that there are lots of different ways of being successful, and that the world needs people who respond differentially. We have tried to show that if all people responded in like manner that it would not only be a dull world, but that our survival as a species might very well be threatened. We have tried to show that we should tend to value diversity more, and that we should rack our brains for ways of arranging the environment so that people can be more creative, diverse, and productive and still feel good about themselves. One of the places where it is most important to allow for diversity is the school. The school is the place that society has designated for its values to be transmitted. It is the place where we send our very young during and just after their developmental years, the time in their lives when they are learning the most. Yet the school is the place where we do the worst job of teaching people to perform and to evaluate others with the idea of individual differences in mind. Just as in the research cited in this text, there are exceptions. There are some schools doing a marvelous job, but they are distinctly in the minority. Most schools are doing poorly, and that is being rather charitable. There are some that are just horrible. One educator recently pointed out that the public school is the last institution in America where a person can be legally beaten, and that is a fact which does not speak well of our "love" for our children.

It is hoped that many parents will read this text; in fact, the title was chosen to get the attention of parents. The section on individualized instruction is included not just because it fits so well with the eye color research findings, but because we would like to influence parents to inspect schools, classrooms, instructional programs, learning materials, and student evaluation methods. We would also like parents to know that there are alternative methods of having school.

Research has shown that certain people respond to their environment in a reactive way. They are highly sensitive to stimulation and tend not to inhibit responses. These same people are often required to sit quietly while a lecture is in progress and are asked to engage in deliberate and elaborate problem-solving exercises. Now there is nothing inherently wrong with "having school" in this manner, for life is filled with situations in which these kinds of behaviors are required for success. What we feel is an injustice is that in many cases the academic part of school is conducted in this fashion exclusively. The result is that sometimes the classroom becomes an aversive place for students; often this aversiveness extends not only to the classroom in general, but to each of the academic activities therein. The punishment comes in many forms. Students are sometimes unable to withhold disclosing to the entire class a sudden insight or the correct solution to a difficult problem. Often the teacher's need for structure and order prompts his or her disapproval of "blurting out" in class. In the chapter on sociability, we reviewed evidence that suggested that there are optimal levels of stimulus need that vary with eye color. It is possible that some individuals suffer much discomfort if the environment does not produce a good bit of stimulation for them. In most traditional classes, students are evaluated for grading purposes almost exclusively on "paper and pencil" tests that require a preponderance of self-paced behavior. A teacher who has a student with no hands would not say, "All right, everyone take out paper and pencil; we are going to take a test. All answers must be written in complete sentences; no other type of testing will be provided." The high visibility of the child with the individual difference of no hands would serve to restrain any teacher from evaluating the academic progress of that child by paper and pencil test. An alternative method would be designed. But there are individual differences in response styles that are never taken into consideration in the typical classroom. True, they are not as obvious as the one cited above, but they are nonetheless there. The argument that one must be taught or evaluated by a particular method just because there are lots of activities in the world that require paper and pencil performance (to use our example above) falls apart when the individual difference becomes obvious.

But what about the self-paced child; don't school procedures favor such a student? In many ways this is true. However, there are instances when the self-paced student is "punished" by the setting conditions by virtue of his response mode. For example, what about the distraction posed by the reactive youngster who "blurts out" in class. Surely this must affect a deliberate performer's quality of work. There are instances when the self-paced youngster may not respond to the teacher's questions in open classroom as quickly as is expected. The teacher may get the impression that the student needs practice speaking before a group; or that the student is withdrawn or emotionally disturbed; or worse still, that the student is stupid or slow. Such generalizations about the emotional or intellectual state of an individual are the result of the use of one standard or norm by which we daily render judgments about others.

Eye color research may be useful to teachers and parents in a number of ways. First of all, it can help to alert teachers and parents that there are individual differences in the way students respond to their environment, and *that these differences are not necessarily intentional.* Next, eye color is an obvious characteristic that may be observed without the use of elaborate and painstaking psychological examination. By noting the students' typical behavior patterns, or better still, taking the trouble to take frequency checks on certain behaviors, one may determine to what extent a given individual is self-paced or reactive in his response mode. Remember, there are many exceptions to the reactivity framework; eye color is just a highly visible (no humor intended) characteristic one may use as a "rule of thumb" to begin the investigation in determining how one can help an individual learn more efficiently. The next step is the choice of the type of individualized instruction that is best for a given child. For students who are self-paced, the traditional approach is probably biased somewhat in their favor; they also seem to do very well with programmed instruction. However, just because one is a self-paced learner does not mean that he should be confined only to self-paced instructional activities. It is important for all students to have as wide a variety of experiences as is possible without their being punitive or aversive. The idea of nonpunishment is important. What we recommend is that the teacher should know something about a student's response mode not only to provide the proper activities, but also to properly evaluate that child's performance. It would be highly inappropriate for the teacher to evaluate a self-paced individual heavily on how quickly he or she typically responded to quick, short response items in the open classroom. It would be equally inappropriate to evaluate the highly reactive youngster almost exclusively on the degree of elaboration in the student's paper and pencil essay questions.

These ideas are all well and good, you say, but how do I as a parent see that some of these things are carried out in the classroom? One way is to become informed on the different kinds of individualized programs now available commercially. One does not have to be a professional educator to find out about the various systems available to schools. (See the Appendix for a list of learning packet sources.) Usually, the next step is to request to inspect your school facility, its curriculum, its methods and materials, and to talk with members of the faculty and administration. This may be done through your parent-teacher organization, through a local civic organization which serves the public schools, or as a private citizen. Incidentally, as a private citizen you have a right to know about your school. Schools are everybody's business. After you have become informed of the alternative methods of instruction available and have acquainted yourself with your local school system, its personnel, and its methods and procedures, it may be that you will be happy with what you see. But in case you aren't, the next step is in order—that of trying to institute change.

Causing change in public school systems is difficult, particularly if you look for things to happen right away, or if you go about your efforts in a way that is threatening to your school personnel. After all, they are the professionals and you are just a layman; you are not supposed to know anything is wrong—much less how to correct it.

School personnel *are* overworked, and they *are* a bit harried at all the criticism that has been leveled at them in the past few years. Therefore, one of the best ways to institute change is to present yourself to one of the organizations that provide tutoring services, social services, and so on, and ask for something to do. Get inside the organization, so to speak. And if there is no organization, form one. If you present yourself carefully so as not to threaten teachers and administration, if you come well informed and with a sincere desire to help and not just to criticize you may be very surprised at the warm welcome you receive. One of the most important moves in the history of American education is afoot; and although its progress has been slow, education is destined to change. As a concerned citizen and parent, you have the opportunity to be a part of that exciting happening.

What should one look for as tools for evaluating school programs?

In *A Climate for Individuality* (1965), a joint statement prepared by the American Association of School Administrators, Association for Supervision and Curriculum Development, National Association of Secondary-School Principals, and Department of Rural Education— all departments of the National Education Association— several things are mentioned as necessary for providing the climate of individuality of instructions. We will abridge this article and mention some of the factors that they felt to be important. For ease of discussion, we will list them in several categories.

PHYSICAL FACILITIES

Can room sizes be easily changed? Do the classrooms allow for small group gatherings? Are the facilities available for independent study? Are language labs available (we assume this includes English)? Do the science labs allow for individual projects? Is room available for individual music, speech lessons, or for watching movies and slides? Is there a place for relaxation and informal discussion? Is there room for individualized sports? Is there an instructional center available for teachers with things such as duplicators, typewriters, and photographic equipment?

Let us stop for a moment and consider the schools we have worked in or are working in. How many of them meet more than one or two of these criteria? Certainly some arguments can be made for the spacing concept for small groups and so forth, but how much of this was actually planned? Even worse, how many of you have ever seen "open classroom" schools with the nifty movable walls and exotic equipment being used as if it were a "traditional" school, and the new gimmicks were nothing more than a nuisance?

Obviously, very few schools fit this pattern. No matter what level of school we consider, or in what income bracket the students fall, it is not until one enters graduate school that one has access to many of the things these highly respected associations feel are necessary. We are not now meeting the physical needs of individualizing instruction and it is not likely that we will do so unless there is some sort of drastic change.

Don't be discouraged, we aren't (at least not much). Improvisation and desire to change are a lot more important than the building you work in. Some of the best work the world has ever seen has come from places that were certainly not physically conducive to learning. The most drastic example that we can think of is the reoganization of thinking that resulted in the world-famous works of Victor Frankl. His learning experiences took place in Nazi death camps during World War II. Obviously, good old County High was to be a lot more physically conducive to learning and teaching.

ORGANIZATIONAL PATTERN

Are teachers allowed time at the beginning of each year for planning? Is time allotted for teachers to complete their record keeping, allowing them to complete what they have learned about individuals? Does the schedule of each teacher allow individual conferences with students? Is the overall schedule flexible enough to allow for the change in time lengths of class periods? Does the general program distinguish between those learning activities that can be carried out in small groups and those that must be administered via large groups? Are the pupil-teacher ratios low enough that a teacher can work with each individual separately and develop understanding of him? Is the continuous growth concept of learning allowed for in the school's organization patterns?

Well, considering the general run of schools, it seems that once again we do not meet these organizational needs for individualized instruction. There seems to be very little a teacher can do without going above and beyond the call of duty to meet these needs. If the school program as a whole is not dedicated to meeting these needs you are going to be in for a long row to hoe, indeed. It is our hope that, ultimately, many of these problems stemming from organization can be solved by the resourceful and dedicated teacher. The only real answer we have is to work hard. After all, we all, in some way, modify the future of hundreds of lives during our contact with our students.

THE TEACHER

We have subsumed several of the various committees' questions into one, the answer to which is most crucial for individualized instruction: Is the teacher committed to doing all that is possible to help meet the needs of his students? You may call this dedication if you wish, the end result is the same. Without the dedication and caring for students on an individual basis, any program is doomed to failure.

CURRICULUM

Is the educational program flexible enough to allow the teacher to maximize the learning experiences for each individual? Are the learning experiences organized in a manner that will insure a reasonable amount of success for each individual? *Is the pattern of the curriculum determined only after consistent identification of differences in pupil characteristics, and subject to revision whenever such a revision or change is necessary?* Are provisions made for exceptional children, whether they are handicapped in some fashion or are extremely gifted? Are groupings kept in such fashion to allow for flexibility concerning subjects, interests, and aptitudes?

Most of these questions are directed in a very straightforward fashion to the planning that has gone into the curriculum for a particular school. All schools have different needs since they all serve different people. Costs, logistics, staffing, and other factors are all evaluated in terms of curriculum before the decisions are made about the type of curriculum to be provided in any system. Our contention, as you might have inferred by the italicized question, is that not enough attention is given to the indentification of individual and group differences. As the entire tone of our book suggests, eye color and, more precisely, reactive and self-paced styles of responding must be considered before any cogent decisions can be made. Our contention goes one step further. We believe, and feel we have shown evidence for this belief, that no form of individualized instructional scheme can work at its potential rate of effectiveness without first examining the reactive/self-paced dichotomy and developing materials suited for individuals falling in both modes of responding.

INSTRUCTIONAL METHODS AND MATERIALS

The questions that the various groups ask in this section, we feel, can also be subsumed into a shorter, more inclusive question. Are the methods and materials truly individualized? In answering this question a lot of things need to be considered; for example, ability, age, interest, need, cognitive style, and response set. Unfortunately, we feel that all too often, very little is done to try to individualize methods or materials in terms of the cognitive style or the mode of responding of the individuals, be it reactive or self-paced.

We purposely did not discuss many of the questions raised by this report, because they deal with political machinery and conditions far beyond the scope of this book. We have examined the few that we did in order to raise some questions about individualized instruction, specifically that question concerning the mode of responding of an individual. Considering all the differences, and with the purpose of providing more rationale feel no qualms in stating that individualization cannot be complete without measurement and exploitation of the reactive/self-paced dichotomy.

The questions that the various groups ask in this strayed somewhat from the focus of the book. Not so; we began the presentation of eye color research with the thought of illustrating the nature of individual differences, and with the purpose of providing more rational for the use of individualized instructional programs and evaluation of students. Behavioral psychology has done a great deal for education and the student in this regard;

however, behavior modification has suffered from a bad case of "poor image" because some of its critics unjustly have characterized it as an "inhuman" approach to instruction. Nothing could be further from the truth. Nothing dehumanizes more or lowers the value of an individual's rights than to be taught as a faceless component of a mass and to be evaluated by a single measuring stick. But you should not take our word for that, or anybody else's, for that matter; see for yourself. Good luck.

PSYCHOLOGICAL COUNSELING

Behavioral reactivity as defined in the eye color research literature may be very important to counseling, both from the standpoint of the selection of counselor candidates and from the standpoint of which counselor-to-be shall receive which kind of training.

The evidence reviewed thus far suggests that dark-eyed and female persons are much more sensitive to environmental stimuli, that they are more responsive behaviorally, and that they generally make better models. It seems that these skills are paramount for at least one type of counselor, the personal-social counselor. Couselors-to-be are now selected by a wide range of criteria, ranging from grade point average and scores on the GRE (Graduate Record Exam) to how the applicant impresses the admissions committee. Current figures are not available, but it is a safe guess that the majority of people admitted to counselor training programs are males. Although the women's liberation movement has helped to remove some of the bias from selection procedures for many occupations and professions, it is our guess that the preponderance of new counselors are men. It is not being suggested that no more males be selected for counselor training, but that we begin to define behaviorally what we wish counselors to be able to do, and further, to stipulate some of the skills in the affective domain that we wish them to display at the outset. It seems that high sensitivity as is typically found in the reactive individual might be one of those skills. The research on creativity and eye color, particularly that which indicates that novel responses tend to come from dark-eyed individuals and that high flexibility is highly correlated with behavioral responsivity, shows that we should pay more attention to factors other than grade point average and GRE scores in the selection of candidates. How eye color and sex are related to effective counseling is indeed a fertile area for graduate research.

We believe we have shown rather conclusively that persons vary in their response styles by eye color and sex; and if that is so, we should expect these persons to respond differentially to various therapeutic treatments. While one person might respond quite well to modeling therapy or to behavior modification (perhaps a reactive type), another might respond better to a program of bibliotherapy, or to a self-imposed behavior mod program (the self-paced individual). Very often the process by which a given person is assigned to a particular counselor is a random one, or persons are assigned to counselors by the order of their presentation at the clinic or counseling center. Perhaps some research should be done on matching counselor and client by eye color. The therapy procedure used with a client is more often than not a function of the kind of training the counselor has had. For instance, a given client might respond much better to a modeling procedure, but that skill might not be in the repertoire of the counselor to which the client is assigned. Perhaps we need to carry out some research on various therapeutic procedures by eye color of counselor. We are fairly certain that the schools of counselor training rarely offer programs of equal emphasis on all types of counseling methods. Therefore, there must be many people who have adopted a particular method largely because it happened to be the one being "pushed" at their school. It is not reasonable to assume that counselors-to-be are able to pick the graduate schools that offer the type of training that suits their inherited response styles. It is the responsibility of counselor training facilities to research this problem. It is generally known by both the faculty and the trainee when the program does not suit an individual, but this is only after the program is well under way and there is a commitment to the trainee. It is difficult to "wash out" a candidate after both the faculty and individual have spent a lot of time and effort in training. There should be better methods of determining ahead of time whether a candidate's response style is suited to the school's method. We believe that eye color research will help in that regard.

THE ARTS

The arts is an area in which the basis for our speculations is better founded than in the previous section. For instance, we know that color and form perception is affected by the eye color of the perceiver. A number of research studies have been done that indicate the competing functions of the eye are relative in strength, depending on the darkness of the eye pigmentation. We also know the dark-eyed persons are much more attentive to color, particularly at the "short" end of the spectrum. We have good reason to believe that light-eyed persons are generally better perceivers of form than people with dark eyes. The kind of information furnished in Worthy's book and the follow-up research presented in chapter 6 of this volume, should provide sufficient impetus for educators to begin to arrange early childhood educational experiences for the *early developmental* enhancement of natural abilities in *each* child. It is not meant that students should be "tracked" into one type of art activity from early childhood; rather, it is meant that education should become sensitive to the natural abilities (and modes of learning) during the early years so that success experiences can be provided early—but not to the exclusion of anyone from any activity based on eye color. For example, it may be that eye color research will contribute to the diagnostic and "readiness" measurement techniques we now have, so that we do a better job of predicting where a person's greatest abilities lie. The predictions might then provide the basis for helping each child maximize his potential. This is in contrast to what is typically now done. We assume that every child needs an equal amount of instruction time, practice, and so on, in the activities that have been selected for all children to participate in. This kind of procedure recognizes no individual differences. How many potentially skilled painters, for instance, have been "turned off" during their early developmental years simply because we had no methods (except by extended psychological testing not widely available to the general school population) with which to detect their talent.

Music is an extremely interesting area. One of the questions that came to our minds when the eye color research began had to do with musicianship. For example, there are two general types of musicians (no, we don't mean good and bad). One musician is the person who reads music "cold" (having never seen it before) and generally can play the piece quite well. There is another kind of musician who does not read well (sometimes not at all) but plays just as well from a technical standpoint. Sometimes the latter type of musician is an adept improvisationalist. An improvisationalist is one who not only can play the melody "as it is written," but can improvise or "play around" the music, and in so doing, he improves upon what the composer has written. A good jazz group will *all* be improvising at times, and still the piece does not lose its basic sound and form, as most sensitive listeners will avow. Improvisiation is an example of creativity; specifically, it is made up of the elaboration, the divergence, and the redefinition components of creativity that is being expressed. Are dark-eyed persons better improvisationalists: With all the musicians in a particular musicians' union divided by eye color, are there proportionately more improvisationalists among the dark-eyed musicians or among the light-eyed? Here is another that needs answering: Among all the musicians in a given musicians' union, are there proportionately more nonreaders of music among the light-eyed group or the dark-eyed group? Do improvisationalists learn their musicianship from modeling other improvisationalists, or do they learn it by a self-paced shaping method, as from the neighborhood piano teacher?

Based on what we know about their sensitivity to external stimuli, it is our guess that the best improvisationalists are dark-eyed individuals. It is also our guess that, proportionately speaking, there are more non-music-reading musicians among the dark-eyed population than among the light-eyed. We have absolutely no data upon which to base this other than the Worthy eye color and reactivity schema; however, it does seem to be a reasonable hypothesis.

Research on musicians and their particular musical "styles" could be extremely valuable information for music instructional modules, particularly those that deal with teaching young children music. This is an area in which it is strongly urged that someone carry out some rigorous experimental research. The same is true for sculpting and the dance. We suspect that highly successful sculptors and ballet performers are primarily light-eyed. Both require deliberate, self-paced skills, and a high tolerance for withholding responses—all associated with light-eyed persons. Touring companies, both from the United States and other countries would make good experimental populations to investigate the question on eye color and dancing skills.

SPORTS AND RECREATION

We are not absolutely certain about what the future would be like. Our best guess is that as the world population grows and as technology becomes more and more important in all of our lives, there will be several effects. The one we wish to point out is the one concerning developing technology. From all we can ascertain at the present time, increasing numbers of people will be displaced from routine jobs, and more and more leisure time will be available to most of us. What all this means is that people will need to put increasing emphasis in their educational years on learning how to be creative and how to "play." The traditional emphasis in educational settings has been on how to "work"; now we must somehow reverse that emphasis. But how does eye color fit with what we expect of the future?

Worthy's research has shown that successful persons in professional sports have eye colors that are predictable by the nature of the sport or the position played within the sport. This kind of "actuarial" information is very important for the design of early childhood and elementary school physical education and recreation programs. It is now the case that the physical education and recreation programs that are offered are not wide enough in scope and are offered to students without any regard for their response styles. So much is learned so quickly during the developmental years that it is easy for a given youngster to miss completely certain activities. This is sometimes true because of the random-like procedures of assigning students to programs, even when a wide range of programs is offered. In other instances, programs that are chosen for the curriculum are not designed to enhance both self-paced and reactive skills. In either event, a would-be Olympic ice skater, a champion swimmer, a skilled gymnast, and so on, may be lost forever to his or her potential simply because we either do not know how to predict what individuals are capable of excelling at, or we do not offer the activities at all. It is our belief that eye color research has a good deal to tell us about the formulation of programs and the cycling of students through such programs. Eye color is a highly visible criterion that will substantially aid in the assessment of early childhood behavioral proclivities, and not just for the enhancement of a particular set of skills such as "reactive activities," but for the early training in skills that are predicted to be low. We have maintained all along that environmental manipulation holds much more promise for the changing of behaviors than knowledge of inherited characteristics. But that does not prevent us from looking at response styles very early during the developmental years with an eye to drastic improvement that might not ever come about later, even with the most intense training. Developmental psychology has been telling us for years that certain things are learned at certain times, and in a certain order. If we are able to refine that knowledge to the extent that we can predict by eye color where a given individual's strengths and weaknesses lie, then we will be in a much better position to provide remediation training. Since the notion of "strengths and weaknesses" is mentioned above, perhaps we should hasten to add that we are not implying that any *one* response tendency is inherently superior or inferior. Our self-paced and reactive skills seem generally to be ipsative; that is, for any given individual who happens to be "good" at self-paced activities, the probability is that the same individual will not be as "good" at reactive activities. Of course, there are those among us who are "good" at both. The point of our educational argument is that it is better to be able to swim a little than not to be able to swim at all. The wider the range of activities a person can perform to some degree or another, the more recreational alternatives are open to him eventually. And if we know at the onset which ones he is likely to be able to do best, we can take the trouble to offer experiences and training in those with the idea of maximizing his proficiencies. We can also spend some time in his weak areas so that he will have "some" competencies here rather than "none."

BUSINESS AND INDUSTRY

In a landmark decision between the Duke Power Company and plaintiff Griggs, the Supreme Court handed down a ruling that is destined to affect radically the hiring and promotion practices in industry and business. In essence, the decision stated that hiring and job promotion can no longer be based on IQ or psychological testing unless the validity of the test can not only identify successful applicants, but also unsuccessful applicants. The eventual result will be that virtually all companies, especially those engaging in interstate commerce, will be required to develop tests or measures of specific job performances. Successful applicants will be required to make a passing score on test items which measure the specific activities that are required for success on a particular job. No longer will the number of years of formal education or scores on IQ or other psychological exams that alledgedly measure one's ability to perform a job serve as candidate selection devices. The measures must now be direct. The decision was aimed at stopping discriminatory practices in hiring and promoting personnel. But its effect reaches far beyond that of personnel practices. For one thing, it may affect all hiring and promotional policies in all organizations such that all jobs might eventually have to be behaviorally stated, with job selection criteria that are directly measurable. What do sex and eye color have to do with the Griggs versus the Duke Power Company decision?

If our interpretation of the decision is correct, companies may eventually have to write highly specific tests to fit each job in their particular industries. If this is done, and if the tests measure only what they are intended to measure, then one need not be concerned beyond that clear and direct measure. While this is true, it is believed that if employers pay attention to the findings on the eye color phenomenon, screening of candidates can become a more efficient process. It has been shown that certain individuals clearly have an advantage over others in terms of the quality of their performances on certain tasks. Many jobs in industry are clearly reactive in nature; others are self-paced. There are others that are a mixture of the two. Where the problems will lie perhaps will be the way in which a given company requires that a job be performed. For instance, there are many jobs that may be done with equal success, but performed in either a self-paced or reactive manner. The issue involves test construction. It may very well be that there will have to be two or more different types of tests for each position to guarantee that no person has been discriminated against in the job selection process. The fundamental legal issue that may arise has to do with whether a company will have the right to determine the nature of the execution of the job process. If a company writes, for instance, a specific job performance outlining not only the terminal criterion for success, but the *process* by which the end result is reached; and if the job test is designed in accordance with the aforestated; if various response styles are not taken into consideration, the test could be determined to be discriminatory.

Take as example the postion of a particular supervisory job. It may be that the company manual will describe certain end conditions for successful supervisors' employees to attain. They might include such things as production level, quality control, employee absenteeism levels, and a host of other factors. The manual may then describe certain self-paced behaviors that the supervisor must engage in to reach these particular goals. However, it may be that it can be shown that a person can arrive at the same success criteria by a reactive method of supervising. Some supervisors are highly successful in using a highly specific, well-planned, and well-executed procedural plan. There are other supervisors who seem to be crisis-oriented in their supervision; that is, they seem to be able to respond appropriately to whatever stimuli are presented. Here we have an example of more than one way of doing a job. It seems plausible that a company would likely be most concerned about the terminal measures used to determine a given person's success on a particular job. But there is the danger the the test for such a job will include particular activities required of the person, activities that are seen by the company as the *only* ones leading to success. The principle being applied here is the same as the one applied in the student evaluation procedure in the school or academic setting. The question is: Is there only *one* way to accomplish the goal without changing the nature of the goal? In many cases, we believe the assumption is just that, there there *is* only one way to achieve a particular goal. We believe the assumption to be erroneous in many instances.

In the personnel selection process in industry, and in virtually any other situation in which a person's performance is to be evaluated, it is our belief that there should not necessarily be a unitary *process* of arriving at an objective (which is held up as the single success criterion). This does not mean that the end result should be modified to suit the response mode of an individual. If the job requires that 843 widgets be boxed in an hour, then that criterion should be maintained. But there may be many ways that 843 widgets can get boxed, and the testing process should allow for them.

It is maintained, therefore, that eye color research can work to the advantage of both employer and employee. By attending to the eye color findings and extending the research to industrial settings, much more knowledge may be gained about the nature of behavior in highly specific situations. Employers may do a much better job of placement than before. Employees may not only be able to find positions that are much more to their liking by virtue of being placed in jobs that allow for their response styles, but at the same time be guaranteed that they are not being bypassed in the awarding of promotions.

There are a number of industries that might profit from further exploration of the Worthy eye color theory. Insurance is probably the one that is most likely to profit the most from reactivity research. Actuarial tables are built from the complex analysis of a large number of factors that affect the payment of claims. The factors include disease hereditability, age, sex, job or profession, personal history, marital status, educational level, and a host of others. The insurance industry presently does not know the relationship between eye color and suicide, automobile accidents, various disease entities, emotional instability, addictive behavior, interpersonal aggression, age at death, and many other factors of interest to the actuaries. Since the ability to inhibit a response or the ability to respond appropriately to new stimuli seem critical to the building of actuarial tables, insurance companies should certainly begin to research eye color.

Most insurance companies gather physical data, including eye color, on most of their applicants. It would be a fairly simple process, since most of them are computerized, to retrieve the necessary information to turn a number of highly important studies. There are a number of designs that immediately come to mind. It would be very interesting, and perhaps financially rewarding, for insurance companies to look at eye color and longevity. There is evidence suggestive that light-eyed people simply live longer. It is not known why, but this is certainly a factor that insurance companies could verify in their own research and adjust their actuarial rates accordingly. We would certainly expect that response inhibition is related to automobile accidents in a significant way. It is not known whether highly reactive people are better at avoiding accidents and therefore are better risks, or if they act on impulse and have a great many more accidents. It may be that neither is the case; however, based on the evidence on eye color and behavior reviewed thus far, reactivity is suspected to affect automobile accidents significantly.

Obesity is a factor that insurance companies are interested in, for it is related to a number of ailments of which heart trouble is the most prominent. There are two lines of thought concerning eye color research that suggest that dark-eyed people are more inclined to obesity than are light-eyed. A surface explanation is simply that highly reactive people would tend to, by definition, be poorer at inhibiting the response of eating, and hence, be fatter. A more in-depth explanation is that the amount of light entering the eye probably affects the metabolic rate of individuals. At any rate, the relationship between eye color and obesity is shown in Worthy's text, and insurance companies would be well advised to look into it. An interesting aside is that knowledge of the relationship between eye color and obesity may eventually bring that very troublesome health problem under control. It is not likely that a procedure as simple as refracting the light with nonprescription contact lenses would lower one's tendency to eat or engage in the addictive behaviors, but wouldn't it be remarkable if that turned out to be the remedy? It is more likely that medical researchers may find an endocrine link that is affected by light reception, and which may be correctable by medication, leading to a "cure" for obesity.

These are the factors that insurance companies should be interested in, the the capability they have for researching them in relation to eye color is intriguing indeed. It is our hope that someone will be prompted to do the investigation.

MARITAL RELATIONS AND
SEXUAL ADJUSTMENT

There is perhaps no area of human interaction that has as much written about it with apparently so little avail as marriage and sexual adjustment. It is for that reason that this area is mentioned with great trepidation. Most of those who have written on the topic before us have had so little of value to say that the worst that can happen is that we will contribute to the general lack of knowledge. And at best some research might be triggered that would aid persons to make better marriage partner selections, and once the selections are made, to understand the behavior of the chosen mate.

Perhaps we should begin by saying that unhappy marriages, like other major human behavior problems, are multifactor in cause, and we do not wish to lead anyone to believe (or even suspect) that we have the answers to marital strife. It is also fair to point out that sexual incompatibility, while often mentioned as a major cause for divorce, is probably way down the list in actual reasons for divorce.

In addition to sexual problems or sexual incompatibility, what are the main causes for divorce? Many have difficulties that are irreconcilable over how money is to be spent; some divorce over child-rearing practices; some people are unable to agree as to whether to have children in the first place. Although less frequently than in the past, religious differences are cited as a major source of marital disagreement. The list of causes is virtually endless, but they all are traceable to behavioral differences between two people.

How is behavioral reactivity possibly related to marital happiness? We might first consider the behavioral differences between two people with respect to the disposition of their income. It is not uncommon in counseling with young marrieds, especially, that disagreement about how money is to be spent is a primary source of disharmony. Often one partner is prone to impulse buyings, plunges the couple deeply into debt without evaluating their capability to pay, and in the eyes of the other partner, is generally "irresponsible" from a fiscal standpoint. The self-paced partner, on the other hand, is accused of being unable to make a decision, much too worried about the future (which might be nonexistent), too conservative, and generally disagreeable and argumentative over finances. In talking with such couples their differences in response styles are so obvious that one is prompted to wonder how they ever got together initially.

Couples who disagree about child management very often display the self-paced/reactive behavior dichotomy. In the past, sociological arguments have been presented to explain parental differences in ideas about child discipline. The authoritarian versus the democratic method has been widely discussed by sociologists, and both are generally attributed to the child-rearing practices under which the parent was reared. Although it is a certainty that how we treat our children is a modeled behavior, and that the most likely role model is that of our own parents, it is not a certainty that all people model their parents. Often other models are emulated, and sometimes the behavior of people changes radically over time as a result of experiences. It would be foolish to presume that eye color and its attendant general response mode is the culprit in parental disagreement over child management. On the other hand, it could be a significant contributing factor. At this point we do not really know. Our best guess is that if eye color were controlled for in an experimental design, along with sociological and modeling factors, that eye color would significantly reduce the error term.

Although it is the authors' belief that sexual incompatibility is *not* a major source of divorce or marital unhappiness in most instances, it is important enough to investigate in a study which controls for eye color. We are thinking of a very serious and highly controlled type of physiological study of the kind executed by Masters and Johnson. It may well be that the charges such as "lack of spontaneity," "wants sex too often and at inconvenient times," "has weird sexual practices," "is too conventional," may be related to the response modality differences of self-paced and reactive persons. If we are to believe the research findings that the ability of an individual to inhibit a response is a *general* behavioral style, and that it is related to endocrine functions, then we must believe that eye color is related to the physiology of sexual functioning. If eye color research is generated in the area of sexual behavior and significant results are obtained, it may be that the knowledge will not alter marriage or sex partner choice appreciably. The subject area, however, remains an interesting and sometimes titillating topic of inquiry.

EYE COLOR FOR EDUCATIONAL
SCREENING AND DIAGNOSIS

Educators have long sought for highly visible and easily obtainable diagnostic variables to aid in the instructional process. But alas, they are not so easily found. As of this writing a large research project is underway which may assist in the screening, if not preliminary diagnosis, of certain learning disabilities in children. The project covers fourteen school systems with a total student population of approximately twelve thousand. The project was initiated after the analysis of a pilot project designed in part to test the feasibility and practicality of proceeding with eye color research on all the categories of exceptional children—including the gifted.

The pilot project was conducted in a small southern city with a school population of about thirty-five hundred students. A survey instrument was designed to gather a number of descriptive and performance characteristics on students that had been tentatively identified as falling in one or more of the ten areas of student exceptionality. Among the physical characteristics listed was eye color. The psychological characteristics were to be obtained from a series of response items or statements on each personality category. For instance, one of the categories was "creative or gifted"; the category was made up of ten statements known to be associated with creative or gifted students. The teachers were asked to rate each of their students on each of the ten stems or statements, using a five-point scale. There were a number of other personality categories listed, including the aggressive child, the withdrawn child, the child leader, the emotionally disturbed child, and so on. In addition, children were rated on the five major areas of learning disabilities. There were from seven to ten categories under each of the five areas, with each category composed of six to ten behavioral statements or stems. Again, the teachers were asked to rate each of the students on each of the stems. The responses were found to yield results that correlated very highly with findings on psychological and diagnostic tests. The findings were also highly correlated with eye color; so much so that a more comprehensive study was undertaken, involving the larger student population over all the school systems, and with the use of control groups.

The preliminary findings of the pilot study were consistent with what one would expect from previous studies. Among the students designated as having one or more learning disabilities, it was found that eye color followed the same pattern as in the Worthy human and animal studies. The population, which was a highly restricted one, still yielded some startling findings. In the studies cited earlier in this test, it was shown that the most creative (reactive creativity) persons were dark-eyed females, dark-eyed males, light-eyed females, and light-eyed males, in that order. The same held true for the restricted population. When the personality category of aggressive child was examined by eye color, the most aggressive children followed the same trend. Studies in the past have shown males to be represented in a much higher proportion than the sex composition of the population universe would predict. This was not found in the pilot study. Males and females were apportioned about as the general population would predict. However, the proportion of dark-eyed aggressives was significantly higher than that of the light-eyed aggressives. The examination of the withdrawn category also followed the Worthy predictions. The proportion of light-eyed persons displaying withdrawn behaviors was significantly greater than the dark-eyed group. Since the study was a preliminary one and did not have control groups, it is believed that the results which will be obtained from the larger population will be even more significant.

If the findings hold up, and there is no reason to doubt that they will not, it is believed that a good foundation can be laid for instituting individualized instructional programs based on response style, with the initial predictor as to who will receive which program being student eye color. Perhaps some inroads can be made on the practice of labeling students "aggressive," "withdrawn," and so forth, and more time devoted to the formulation of individualized programs based on response modes.

As in all research cited, the reader is urged not to make decisions about *individuals* based on eye color; we are reporting general trends by eye color groups. We also add the exhortation to replicate the research designs.

There are many other areas upon which we would like to speculate. Perhaps we have omitted an area that is of great interest to you. If so, we implore you not to let your curiosity go unsatisfied—research it.

Appendix

UNIPAC	Teachers UNIPAC exchange W.B. Field and Associates Box 332 Miami-Kendall, Florida 33156
LAP	Continuous Progress Program Hughson Union High School Box 98 Hughson, California 95326
PLAN	Westinghouse Learning Corporation 2680 Hanover Street Palo Alto, California 94304
OMAPAC	Omaha Public School System 3902 Davenport Street Omaha, Nebraska 68131

IPI   University of Pittsburgh
Learning Research and
Development Center
Pittsburgh, Pennsylvania

Bibliography

Ackerman, J.M. *Operant Conditioning Techniques for the Classroom Teacher.* Glenview, Ill.: Scott, Foresman, 1972.

Bandura, A., Ross, D., & Ross, S.A. Imitation of film-mediated aggressive models. *Journal of Abnormal and Social Psychology,* 1963, *66,* 3-11.

Barron, F.X. *Creative Ability.* New York: Holt, Rinehart & Winston, 1969.

Bart, R.S., & Schnall, S. Eye color and darkly pigmented basal-cell carcinomas and malignant melanomas. *Archives of Dermatology,* 1973, *107,* n.p.

Berry, J.W. Mueller-Lyer susceptibility: Culture, ecology or race? *International Journal of Psychology,* 1971, *6,* 193-197.

Birdwhistell, R.L. *Kinesics and Context.* Philadelphia: University of Pennsylvania Press, 1970.

Bohm, D.W. Originality in culturally disadvantaged children as a function of nonverbal reinforcement. Unpublished master's thesis, University of Tennessee, 1971.

Bristol, B.K. Some traits of creativity. *Journal of Creative Behavior,* 1971, *5,* 1-6.

Climate for Individuality, A. American Association of School Administrators, Association for Supervision and Curriculum Development, National Association of Secondary School Principals, and Department of Rural Education. Stock No. 021-00590. National Education Association, 1965.

Crosley, A.C. *Creativity and Performance in Industrial Organizations.* London: Tavistock Publications, 1968.

Dacey, J.S. Programmed instruction in creativity and its effects on eighth grade students. (Doctoral dissertation, Cornell University, 1967.) *Dissertation Abstracts International,* 1967, *32,* 2479A.

Dellas, R.M. A comparison of two methods of instructing for creativity. Unpublished doctoral dissertation, University of Wisconsin, 1970.

Drea, P.S. Development in creativity. *Journal of Counseling Psychology,* 1972, *19,* 53-57.

Dunn, J.R. & Lupfer, M. A comparison of black and white boys' performances on self-paced and reactive sports activities. Paper presented at the meeting of the Southeastern Psychological Association, Atlanta, April 1972.

Edwards, T. M. Creativity and reflectivity testing: The effects of environment on performance. Unpublished doctoral dissertation, Harvard University, 1970.

Elie, M.T. A comparative study of middle school and junior high school students in terms of socio-emotional problems, self-concept of ability to learn, creative thinking ability, and physical fitness and health. (Doctoral dissertation, Michigan State University, 1971.) *Dissertation Abstracts International,* 1971, *31,* 5696A.

Flanagan J.C. *Flanagan Aptitude Classification Test No. 18—Ingenuity.* Chicago: Science Research Associates, 1967.

Gary, A.L. Color and form perception in relation to self-paced and reactive behavior in children diagnosed as having specific learning disabilities. Unpublished manuscript, 1972.

Gary, A.L., & Glover, J.A. Eye color and sex: Their effect on modeling creativity. Paper presented at the meeting of the Southeastern Psychological Association. Miami, May 1974; Psychotherapy: Theory, Research, and Practice (in press).

Gary, A.L., & Glover, J.A. Eye color and developmental skills. Journal of Psychology July, 1975.

Glover, J.A. Modeling as a technique for inducing creativity. Unpublished doctoral dissertation, University of Tennessee, 1973.

Glover, J.A. & Gary, A.L. Melanin and its association with acquisition and extinction rates of a modeled behavior. Unpublished manuscript, 1974.

Goetz, E.M., & Baer, D.M. Social reinforcement of "creative" blockbuilding by young children. In E.A. Ramp & B.L. Hopkins (Eds), A New Direction for Education: Behavior Analysis. Lawrence: University of Kansas Printing Service, 1972.

Goetz, E.M. & Salmonson, M.M. The effect of general and descriptive reinforcement on "creativity" in easel painting. Paper presented to the Third Annual Conference on Behavior Analysis in Education, Lawrence, Kan., 1972.

Gowan, J.C. Development of the creative individual. Gifted Child Quarterly, 1971, 15, 156-174.

Guilford, J.P. Uses Test. Beverly Hills, Calif.: Sheridan Psychological Services, 1968.

Happy, R., & Collins, J.K. Melanin in the ascending reticular activating system and its possible relationship to autism. Medical Journal of Australia, 1972, 2, 1484-1486.

Harris, D.H., & Simberg, A.L. A-C Test of Creative Ability. San Diego: Education-Industry Service, 1967.

Hinde, R.A. (Ed.). Non-verbal Communication. London: Cambridge University Press, 1972.

Hooper, M. Creativity and the curriculum. Improved College and University Teaching, 1967, 19, 187-188.

Jahoda, G. Retinal pigmentation: Illusion susceptibility and space perception. *International Journal of Psychology*, 1971, *6*, 199-208.

Jones, N.B. (Ed.). *Ethological Studies of Child Behavior*. London: Cambridge University Press, 1972.

Jordan, J.J., III. An investigation of sex, eye darkness and social class differences in perceptual motor and cognitive abilities. Unpublished doctoral dissertation, Georgia State University, 1972.

Karp, E.J. Changes in interpersonal attraction: Effects of physical attractiveness and attitude similarity as measured by the pupilometry technique. Unpublished doctoral dissertation, Emory University, 1972.

Kastein, S., & Trace, B. *The Birth of Language: The Case History of a Non-verbal Child*. Springfield, Ill.: Charles C Thomas, 1966.

Maltzman, I., Bogartz, W., & Breger, L. A procedure for increasing word association originality and its transfer effects. *Journal of Experimental Psychology, 1958, 56*, 392-398.

Markle, A. Effects of eye-color and temporal limitations on self-paced and reactive behaviors. Unpublished doctoral dissertation, Georgia State University, 1972.

Martin, W.D. Encouraging more creative teaching. *Instructor, 1971, 81*, 25-26.

McCalmon, D.H. *Creating Historical Drama*. Carbondale: Southern Illinois Free Press, 1965.

Moore, G.T., & Jay, L.M. *Creative Problem Solving in Architecture*. Berkeley: University of California, Architectural Experimental Laboratory, 1968.

Nichols, J.G. Some effects of testing procedure and divergent thinking. *Child Development*, 1971, 42, 1647-1651.

Parnes, S.J. *Creative Behavior Guidebook*. New York: Scribner's, 1967.

Parnes, S.J. Creativity: Developing human potential. *Journal of Creative Behavior*, 1971, 5, 19-36.

Pollack, R.H., & Silvar, S.D. Magnitude of the Mueller-Lyer illusion in children as a function of pigmentation of the fundus oculi. *Psychonomic Science,* 1967, 7, 159-160.

Pryor, R., Haag, R., & O'Reilly, J. The creative porpoise: Training for novel behavior. *Journal of the Experimental Analysis of Behavior,* 1969, 12, 653-661.

Reese, H.W., & Parnes, S.J. Programming creative behavior. *Child Development,* 1970, 41, 413-423.

Schubert, H.J.P., & Schubert, D.S.P. *Creative Imagination Test.* H.J.P. Schubert, Publisher, 1967.

Scheflen, A.E., & Scheflen, A. *Body Language and Social Order.* Englewood Cliffs, N.J.: Prentice-Hall, 1972.

Skinner, B.F. *About Behaviorism.* New York, Knopf, 1974.

Steinberg, L. Creativity as a character trait: An expanding concept. *California Journal of Instructional Improvement,* 1964, 7 3-9.

Tavris, C. Konrad Lorenz on aggression, homosexuality 1974.

Torrance, E.P. Give the devil his dues. *Gifted Child Quarterly,* 1961, 5, 115-118.

Torrance, E.P. *Guiding Creative Talent.* Englewood Cliffs, N.J.: Prentice-Hall, 1962. (a)

Torrance, E.P. Must creativity be left up to chance? *Gifted Child Quarterly,* 1962, 6, 41-44. (b)

Torrance, E.P. Toward the more humane education of gifted children. *Gifted Child Quarterly,* 1963, 7, 135-145.

Torrance, E.P. *Encouraging Creativity in the Classroom.* Dubuque, Iowa: William C. Brown Co., 1971.

Turknett, R.L. A study of the differential effects of individual versus group reward conditions on the creative productions of elementary school children. Unpublished doctoral dissertation, University of Georgia, 1972.

Walker, J. Creativity and high school climate. Unpublished doctoral dissertation, Syracuse University, 1964.

Warren, T.F. Creative thinking techniques: Four methods of stimulating original ideas in sixth grade students. (Doctoral dissertation, University of Wisconsin, 1971.) *Dissertation Abstracts International,* 1971, *31,* 5863A.

Weiss, B. Food and mood: What you eat may be what's eating you. *Psychology Today,* December 1974.

Weiss, J.M., Glazer, H.I., & Pohorecky, L.A. Neurotransmitters and helplessness: A chemical bridge to depression. *Psychology Today,* December 1974.

Williams, F.E. Teaching for creativity. *Instructor,* 1971, *81,* 42-44.

Wilson, R.E., Merrifield, P.R., & Guilford, J.P. *Utility Test.* Beverly Hills, Calif.: Sheridan Psychological Services, 1968.

Winer, B.L. *Statistical Principles in Experimental Design.* New York: McGraw-Hill, 1971.

Worthy, M. Eye-darkness, race and self-paced athletic performance. Paper presented at the meeting of the Southeastern Psychological Association, Miami, 1971.

Worthy, M. *Eye Color, Race and Sex: Keys to Human and Animal Behavior.* Anderson, S.C.: Droke House/Hallux, 1974.

Worthy, M., & Markle, A. Racial differences in self-paced versus reactive sports activities. *Journal of Personality and Social Psychology,* 1970, 16, 439-443.

Index

About the Authors

John A. Glover is assistant professor of psychology at Tennessee State University, Nashville. He previously was an instructor at the University of Tennessee and was the director of psychological services and counseling at Friendsville Academy. He received his B.S. at Memphis State University and his M.A. at Tennessee Technological University. He obtained his Ed.D. at the University of Tennessee in Knoxville.

Dr. Glover is the senior author of *Behavior Modification: Enhancing Creative and Other Good Behaviors* (1975) with Dr. Gary. He is the author of *Behavior Modification: An Empirical Approach to Self-Discipline* and *A Parent's Guide to Intelligence testing: Developing Your Child's Intellectual Abilities.* Articles in various areas of psychology written by Dr. Glover have appeared in the *Journal of Psychology; Psychological Record; Journal of Genetic Psychology; Journal of Applied Behavior Analysis; Psychotherapy: Theory, Research and Practice;* and other psychological and educational journals.

Albert L. Gary is in private practice as a psychologist and is also a professor of education at the University of Tennessee in Chattanooga. In the past he has been an instructor at Michigan State University and the University of Tennessee in Knoxville. Dr. Gary has served as consulting psychologist to the Southeastern Tennessee Educational Cooperative, mental health consultant and therapist to the North

Mississippi Region Mental Health and Retardation Center, Oxford, and director of seven courses in teacher-training curricula for Hamilton County, Tennessee.

Dr. Gary received his A.B. and M.Ed. at Georgia State University, Atlanta, and his Ed.D. at the University of Tennessee in Knoxville. He has written two previous books: *The Psychic World of Doc Anderson* (1973), and with Dr. Glover, *Behavior Modification: Enhancing Creativity and Other Good Behaviors* (1975). Dr. Gary's writings on self-disclosure and alcohol-drug abuse, absenteeism in industry, new teaching techniques, and eye color have appeared in the *Journal of Personality and Social Psychology; Psychotherapy: Theory, Research and Practice; Personnel Journal; Teaching Exceptional Children; Journal of Applied Behavioral Analysis;* and *Journal of Psychology.*